Rapid Assessment Program

RAP
Working
Papers

A Biological Assessment of the Lakekamu Basin, Papua New Guinea

CONSERVATION INTERNATIONAL

RAP Working Papers are published by:

Conservation International

Department of Conservation Biology

2501 M Street, NW, Suite 200

Washington, DC 20037

USA

202-429-5660

202-887-0193 fax

www.conservation.org

Editor:

Andrew L. Mack

Assistant Editors: Jed Murdoch and Arlo H. Hemphill

Design: Catalone Design Co.

Maps: Jed Murdoch and Arlo H. Hemphill

Cover Photographs: Bruce Beehler and Charles G. Burg

ISBN 1-881173-20-8

This publication has been funded by CI-USAID Cooperative Agreement #PCE-5554-A-00-4028-00

TABLE OF CONTENTS

PARTICIPANTS

Gerald Allen, Ph. D. (ichthyology)
West Australia Museum
Francis St.
Perth, WA 6000 AUSTRALIA
fax: 09-328-8686
email: fish@multiline.com.au

Allen Allison, Ph. D. (herpetology)
Bernice P. Bishop Museum
1525 Bishop St.
Honolulu, HI 96817 USA
fax: 808-847-8252
email: allison@hawaii.edu

Solomon Balagawi (mammalogy)
c/o Student Services
Department of Biology
UPNG, P.O. Box 320
University, N.C.D., PNG

David P. Bickford (herpetology)
University of Miami
Department of Biology
P.O. Box 249118
Coral Gables, FL 33124 USA
fax: 305-284-3039
email: bickford@fig.cox.miami.edu

Paul Igag (ornithology)
P.O. Box 1126
Madang, Madang Province, PNG

Rodney T. Ipu (mammalogy)
c/o Lulpi Yangun
P.O. Box 209
Boroko N. C. D. PNG

Miller Kawanamo (entomology)
c/o Student Services
Department of Biology
UPNG, PO Box 320
University N. C. D., PNG

Stuart Kirsch, Ph.D. (anthropology)
Department of Anthropology
University of Michigan
500 S. State St.
Ann Arbor, MI 48019 USA
fax: 313-763-6077
email: skirsch@umich.edu

Joel Kulang (botany)
Forest Research Institute
P.O. Box 314
Lae, Morobe PNG
fax: 675 472-4357

Andrew L. Mack, Ph. D. (ornithology)
Conservation International
2501 M St., N.W.
Washington, D.C. 20037 USA
Mailing address:
P.O. Box 15
Weikert, PA 17885 USA
fax: 717-922-1152
email: a.mack@conservation.org

Kurt F. Merg (coordinator)
Department of Botany
Washington State University
Pullman, WA 99164

Tau Teeray Raga (mammalogy)
c/o Paul Bina, Pasuwe Ltd.
P.O. Box 6124
Boroko N. C. D., PNG

Alexandra Reich (botany)
Bodega Bay Marine Laboratory
P.O. Box 247
Bodega Bay, CA 94923 USA
email: areich@ucdavis.edu

Stephen Richards, Ph. D. (entomology, herpetology)
James Cook University
Department of Zoology
fax: 61-77-251570
Townsville, QLD 4811 AUSTRALIA
email: stephen.richards@jcu.edu.au

Roy R. Snelling (entomology)
Los Angeles County Museum of Natural History
Department of Entomology
900 Exposition Boulevard
Los Angeles, CA 90007 USA
fax: 213-747-0204
email: antmanrs@lam.mus.ca.us

Wayne Takeuchi, Ph. D. (botany)
Forest Research Institute
P.O. Box 314
Lae, Morobe PNG
fax: 675 472-4357

Geordie Torr (entomology, herpetology)
Department of Zoology
James Cook University
Townsville, QLD 4811 AUSTRALIA
fax: 61-77-251570
email: geordie.torr@jcu.edu.au

Debra D. Wright, Ph. D. (mammalogy)
P.O. Box 15
Weikert, PA 17885 USA
fax: 717-922-1152
email: ddwright@ptd.net

ORGANIZATIONAL PROFILES

CONSERVATION INTERNATIONAL

Conservation International (CI) is an international, non-profit organization with headquarters in Washington, DC. CI believes the Earth's natural heritage must be maintained if future generations are to thrive spiritually, culturally, and economically. Our mission is to conserve biological diversity and the ecological processes that support life on Earth and to demonstrate that human societies are able to live harmoniously with nature. Currently, CI has offices in twenty-three countries.

Conservation International
2501 M St., NW, Suite 200
Washington, D.C. 20037 USA
202-429-5660 (phone)
202-887-0193 (fax)
http://www.conservation.org

Conservation International-PNG
P. O. Box 106
Waigani, NCD, PAPUA NEW GUINEA
675-325-4234 (phone/fax)

FOUNDATION FOR PEOPLE AND COMMUNITY DEVELOPMENT

FPCD is based in Port Moresby, Papua New Guinea. It is PNG's largest rural development NGO. Formerly known as Foundation of the Peoples of the South Pacific/PNG (FSP-PNG), FPCD focuses on issues related to village development and the environment in PNG, and is currently active in village theater, educational outreach, eco-forestry, site-sustainable agriculture, literacy training, and sustainable development. FPCD has taken a lead among PNG NGO's in issues of linking conservation and economic development. The Foundation for People and Community Development has been a partner with CI in the Lakekamu Basin since the project's beginning in 1994. FPCD staff manage the Ivimka Research Station and serve as liaisons with local communities in the Basin.

**Foundation for People and
Community Development, Inc.**
P.O. Box 1119
Boroko, NCD, Papua New Guinea
675-325-8470 (phone)
675-325-2670 (fax)

ACKNOWLEDGEMENTS

Many institutions in PNG contributed to the success of the Lakekamu RAP survey. We particularly wish to thank: the Department of Environment and Conservation, the Forest Research Institute (Lae), The Foundation for People and Community Development, the University of Papua New Guinea, and the Wau Ecology Institute. Mount Holyoke College partially supported the fieldwork of Dr. Kirsch. The National Geographic Society supported fieldwork of Steve Richards and Geordie Torr through a grant to Bruce Beehler. We are grateful to the government of PNG for granting long-term research visas for many of the participants.

The RAP survey was a success due to the outstanding contributions of many individuals, some of whom are affiliated with the above mentioned institutions. We deeply appreciate the assistance of: Bruce Beehler, Ilaiah Bigilale, Frank Bonaccorso, Yati Bun, Chuck Burg, Peter Connors, Kippy Demas, Glenda Fabregras, Mike Hedemark, Neville Howcroft, Roger James, Arlyne Johnson, Gai Kula, Michael Laki, Cynthia Mackie, Cosmas Makamet, John Michalski, Thomas Paka, Dan Polhemus, A. Sitapa, John B. Sengo, Lester Seri, Peter F. Stevens, Jørgen Thomsen, and Tim Werner.

The UPNG students that came to the Lakekamu Basin immediately following the RAP assisted by collecting data that are included with this report. Their help is greatly appreciated: Augustina Y.S. Arobaya (UNCEN), Solomon Balagawi, Mark Ero, Kasbeth S. Evei, Phyllis L. Frank, Banak Gamui, Vidiro Gei, Esthel Gombo, Rodney T. Ipu, Miller Kawanamo, Marjorie Kane, Ruth Na'aru, Julius D. Pano, Tau Teeray Raga, Judith Raka, Stewart Serawe, Joice Taufa, and Tanya Zeriga.

The landowners of the Ivimka area and of Tekadu Village, particularly Peter and Tami Uyapango and their families, granted permission for our study on their land and supported the work in many ways. Donald, Waraks, Edward, Joma and Ricki helped with fieldwork.

Financial support for the project was provided under USAID Cooporative Agreement. #PCE-5554-A-00-4028-00

EXECUTIVE SUMMARY

INTRODUCTION

Papua New Guinea's land tenure laws and traditions serve as an outstanding example to a world where indigenous people's land rights have been trampled by governments and commercial interests. Nowhere outside of the Pacific do indigenous people retain so much control over their land. Because of this, development and exploitation must proceed differently in PNG, as must conservation. PNG cannot simply decree a protected area and keep people out of it. The conservation initiative in the Lakekamu Basin is one example of landowners' and conservationists' adaptation to the special requirements of PNG law and tradition.

The distribution of life on planet earth is not uniform, some areas have more than others; PNG is especially endowed in this regard. The island of New Guinea is a unique biogeographic region that is high in species richness and endemism, containing over 5% of earth's biodiversity in just over one half of one percent of the earth's landmass. The people of PNG can take pride in the rich natural heritage that is theirs and take the lead in conservation for neighboring countries to emulate.

A subset of the terrestrial organisms found in southern PNG can also be found in Queensland, Australia. However, Australia only has small fragments of rainforest habitat remaining compared to PNG's extensive, pristine forests. Only 2073 km^2 of wet tropical forest is protected in Australian National Parks—less than the area of the Lakekamu Basin alone! Despite this, tourism in this area (including the Great Barrier Reef) is estimated to generate an economy in excess of A$400,000,000 (roughly equivalent to the export value of all timber removed from PNG in 1993 and more than three times the government's income from logging in 1993). With proper management, there is no reason the richer biological resources of PNG could not attract a larger share of the Australian tourism economy.

The Rapid Assessment Program (RAP) of Conservation International conducted a one-month biological survey in the Lakekamu Basin followed by a one-month training course for University of PNG (UPNG) biology students in late 1996. One goal of these endeavors was to produce biological data useful in furthering the goals of the Lakekamu Conservation Initiative. This report details the findings of a team of leading scientific experts and UPNG student-trainees. These scientists have contributed chapters on the biodiversity (including all vertebrate taxa, some insect taxa, and plants) of the area and a chapter on the socio-cultural history in the Basin.

Few areas anywhere in New Guinea have been so thoroughly surveyed for a variety of taxa, in this case all vertebrates, plants and selected insect families. In all taxa, except birds and mammals that are fairly well known, 35-47 new species and possibly new genera were discovered. The discovery of so many new species and major range extensions for many other species during just one month of survey highlights the urgent need for more biological inventories and taxonomic studies in PNG. The creation of the Ivimka Research Station in the Lakekamu Basin will hopefully catalyze more research on the unique biota of PNG. As the level of basic biological inventory improves in PNG, it becomes possible to begin the necessary task of ecological study—autecology (of particular species), community ecology (of groups of species) and whole ecosystem studies. Armed with such knowledge it will be easier to conserve PNG's rich biological heritage while developing a better economy and quality-of-life for its citizens. Thus, the data presented here are an important step in a process directed toward building a sustainable, healthy environment in PNG.

This report differs from previous RAP reports by including more technical details because this report is hoped to serve as a foundation for future scientific research in the Ivimka Research Station (IRS) and Lakekamu Basin. However, scientific jargon has been minimized. The reader can learn the main points of the report just by reading the executive summary, and turning to the technical reports when clarification is desired. The main points of each taxonomic group are summarized separately, followed by an overall summary and a list of specific conservation recommendations.

TAXONOMIC SUMMARIES

Plants: Part 1 Vegetation Plots

Two one-hectare vegetation plots were established, one in a hill forest in undulating torrain 175-260 m elevation, and one in flat alluvial forest t 110 m elevation. The plots contained 253 species and added eleven families and roughly 174 species to those found on the general survey (part 2). Thus, in a small portion of the Basin in a short time period the RAP survey revealed the presence of over 600 plant species in 130 plant families. Clearly the Lakekamu Basin is a region of exceptional species richness.

The plots are typical for rainforests in terms of stem density and total basal area, with the hill

plot tending to have a somewhat higher stem density than most sites. Both plots had relatively low counts of large trees. Economically valuable timber species were found in the hill plot, but the alluvial forest seems fairly free of exportable timber species.

Soils on both plots were poor with low nutrient content, low cation exchange capacity and high levels of aluminum and iron. Such soils are usually unsuitable for plantation agriculture and vulnerable to serious degradation from large-scale deforestation.

A previous study examined vegetation on three one-hectare plots in the Nagore/Si region of the Lakekamu Basin. Comparison of the RAP data with that study indicates there is substantial variation in floristics within the Basin. Only three plant families ranked in the top ten families in all five plots and eight families were in the top ten of only one of the five plots. Of the dominant 23 species in the Nagore/Si plots, only three occurred at all (as rare individuals, not dominants) in the Ivimka plots. Although the Lakekamu Basin is essentially a single continuous closed forest, the composition of that forest varies greatly from place to place. The quantitative plot data corroborate the general observations of the RAP scientists that the Basin is a heterogeneous mosaic of vegetation and species communities. Successful conservation in the Basin will necessitate conservation of the full array of vegetation types.

More study is clearly needed to assess the floral and community diversity within the Basin. Any assessment of timber value within the Basin will need to examine the array of vegetation types within the Basin. Failure to do so could grossly over- or under-estimate actual timber value. The creation of two marked plots with representative specimens in the nearby collections at The Forest Research Institute (FRI) enhances the attractiveness of the IRS to scientists considering research in PNG.

Plants: Part 2 General Survey

Two broad forest types were surveyed area: alluvial forest and hill forest. The alluvial forest, found on the level bottomland of the Basin, is very species rich and comprised of a heteroge-

The discovery of so many new species and major range extensions during just one month of survey highlights the urgent need for more biological inventories and taxonomic studies in PNG.

neous mosaic of different plant communities at various seral stages. These seral stages derive from disturbances in the forest due to isolated windfall trees, changes in stream and river courses, large storm-caused windfall areas, and old human disturbance (the Bulldog Road). There is little evidence of recent human disturbance other than that associated with the construction of the IRS. There are no strongly dominant species in the alluvial forest.

The hill forest, found on the slopes adjoining the alluvial forest, is less diverse and contains about ten dominant species. There is less evidence of disturbances (like windthrows) in the hill forest than in the alluvial forest, which could be a contributing factor allowing some species to attain dominance. Interestingly, there is an anomalous component of montane species below their normal elevations in the hill forest and some of these species also extend into the lowland alluvial forest.

The botanical survey team recorded about 450 species and morpho-species of vascular plants during the brief period of the survey, highlighting the high species richness of the area. This flora has significant local endemism and many taxa known only from the Papuan (southern PNG) subregion. As this area is rather poorly known, the collections made on the RAP are very important. Five to nine species new to science were collected and several major range extensions were recorded.

Given the many novelties revealed during this quick survey and the surprising range extensions found, continued botanical inventory of the Lakekamu Basin is highly recommended. A solid understanding of plant ecology and distribution in the Basin is the necessary underpinning for expanding the research programs at the IRS. Judging by the data collected on the ground in a limited area and the heterogeneity in the Basin visible from the air, the vegetation of the entire Basin and adjoining hills promises to be extremely exciting. Worldwide there are few pristine wilderness areas so rich in plant diversity.

Insects: Social Hymenoptera

Social Hymenoptera (ants, social bees, and social wasps) were surveyed using a variety of techniques. A total of 281 species were recorded: two genera and four species of bees; five genera and 23 species of wasps; and 64 genera and 254 species of ants. This represents the greatest diversity of ants known from anywhere in the world. This finding alone makes the Lakekamu Basin a location of global significance in terms of biodiversity. It is likely that other insect groups in the Basin will also contain extraordinary species richness.

At least nineteen of the ant and wasp species collected represent undescribed species and one represents a previously undescribed genus. Many species represent major range extensions including first records for mainland New Guinea and PNG. Moreover the survey period was relatively short and did not sample the canopy where many more species can be expected to be found.

Unlike many other insect groups, social Hymenoptera are present year-round and are reasonably conspicuous. This makes them particularly good candidates for monitoring and further study. The thorough baseline data established during the RAP survey and the high diversity found in the Lakekamu Basin make this an ideal location for continued studies of social Hymenoptera.

Insects: Odonata

Dragonflies and damselflies were also surveyed, yielding a total of 34 species. Among these there is at least one, possibly three, undescribed species and possibly a new genus. Taxonomic knowledge of south PNG's Odonata is poor enough that certain identification of just these 34 species is extremely difficult. One species collected appears to be identical to a species previously known only from Misool Island; clearly more surveys are needed in PNG.

We present a checklist and summarize the habitat preferences of each species. These data should help make it easier for future students and entomologists to study odonates in the Lakekamu Basin. As odonates spend their larval stages in aquatic environments, they can serve as excellent bio-indicators of water quality. Future studies can not only expand our knowledge of odonate natural history, but also provide data useful when monitoring water quality.

Fish

The fish fauna of the upper Lakekamu Basin is broadly typical of freshwater localities in New Guinea. It consists of 23 species in 18 genera and 14 families and is dominated by catfishes, rainbowfishes, gobies and gudgeons. The majority of species recorded are distributed widely either across the southern portion of New Guinea or the combined northern Australia-southern New Guinea region. There is probably no endemism in the fish fauna of the Upper Lakekamu Basin.

Two undescribed gobies belonging to the genera *Glossogobius* and *Lentipes* were collected during the survey, and a species of rainbowfish of the genus *Melanotaenia* is also a possible new species. The new *Lentipes* has now been described (*L. watsoni*) and represents a significant find as the genus was previously unknown from New Guinea. Dr. Allen recently discovered a second *Lentipes* species in the vicinity of Jayapura, Irian Jaya. The new species found on this survey probably occurs in other southern drainages in PNG, emphasizing the need for basic biological surveys in New Guinea.

As one moves upstream the number of fish species diminishes. The species total decreases from 19 to 6 over a distance of approximately 8 kilometers. The Sapoi River in the vicinity of the Ivimka Research Station provides an excellent opportunity to study upstream species attenuation. The rivers of the upper Basin are still pristine and essentially uncontaminated by introduced species (one African tilapia was encountered). Such systems should be studied and monitored in order to provide essential baseline data relevant to water quality issues associated with many development practices (logging, mining, plantation agriculture) throughout PNG. Thus continued study of the fishes of Lakekamu Basin should benefit the conservation initiatives there and help clarify a wide range of conservation and development issues throughout the country.

Herpetofauna

Seventy-four species of amphibians and reptiles (amounting to roughly 14.2% of the known herpetofauna of PNG) were recorded from the Lakekamu Basin during the RAP fieldwork. Eleven species of frogs and seven species of reptiles may be new to science.

Standardized, quantitative sampling methods were employed to establish a baseline for comparison with other sites, and to facilitate on-going monitoring of frog populations. The diversity and density of frogs were high relative to other southern lowland sites in PNG. Populations of frogs in the area were healthy and vigorous, making Lakekamu an important site for ongoing monitoring of amphibians, some of which appear to be declining globally.

As a result of the thorough studies conducted during the RAP, Lakekamu now has one of the best-known herpetofaunal assemblages in PNG. This herpetofauna appears to have its main affinities with elements to the west, and includes several species not previously known east of the Kikori and Purari drainages.

Mammals

Twenty-four mammal species representing eight families and 19 genera (12 rodent, 9 bat and 3 marsupial species) were recorded around the IRS during the six week survey period. Of the 24 mammal species recorded, the IUCN Red Data Book lists one as "data deficient," two as "vulnerable" and one as "threatened." Additional sampling using the techniques employed on the survey would not yield many more species; however, there are certainly many more mammal species in the Basin. Potentially 80 species could occur in the Basin based upon the incomplete distribution data available in the literature. Revealing these species will require employment of different techniques (*e.g.*, canopy-netting, hunting, and trapping with different traps and baits) and surveys at different times of year. Further survey work is recommended to target elusive species.

Basic natural history data, such as diet, reproductive cycles, or habitat requirements are lacking

Conservation of the full spectrum of the Basin's 190 species will require conservation of a large expanse of Basin forest.

for most New Guinea mammals. Several novel natural history observations and range extensions were discovered in the course of the survey.

Continued research at the IRS would yield many important new findings. It is essential to learn more of the natural history and ecological requirements of New Guinea's globally unique mammal assemblage. Without such knowledge, predicting the effects of both development and conservation actions would be largely speculative.

Mammals provide an important source of nutrition and play significant traditional roles for many people in PNG. Studies should be made of hunting and consumption of wild mammals by humans. If coupled with natural history data, such studies could help optimize harvest in a sustainable fashion. The combination of four distinct landowner groups, a diverse mammal fauna, and extensive pristine forest in the Lakekamu Basin makes this area an ideal location for investigation of the interactions between mammals and humans.

Birds

No part of PNG is known to have as many bird species within a single, narrow elevation-vegetation zone: 190 species in roughly 150 m elevational span. This represents over 25% of the birds known to occur in PNG. The Basin is home to one third of the species resident in PNG (excluding migrant species and endemics to the Bismarck Archipelago). Four Basin species are listed as "threatened" and six "near-threatened" in the IUCN Red Data Book.

Due to the extensive avian research previously conducted in the Basin and the advanced state of knowledge in ornithology relative to other taxa, the Basin avifauna is among the best-known for any taxa, anywhere in New Guinea. Only seven species were recorded for the first time in the Lakekamu Basin during the RAP. Baseline data were collected for 68 species during point counts and 467 individuals of 51 species by netting and banding. This solid foundation makes the IRS an ideal site for continuing in-depth studies of avian ecology. Like mammals, birds also play an important role in diet and tradition of many people in PNG. Studies of avian ecology and demography

coupled with study of human use and consumption of birds is strongly recommended.

Comparison of RAP survey data with data collected in other parts of the Basin indicates that bird populations are not uniformly distributed across the Basin. Some species are common at one Basin site and rare or absent at another even though there is continuous lowland forest between sites. Apparently, some parts of the Basin are unsuitable for some species while other parts are suitable. The Basin is not a homogenous forest, but a mosaic of different population centers that collectively comprise an outstanding, exceptionally rich ecosystem. Furthermore, some species require large home ranges within the Basin. Conservation of the full spectrum of the Basin's 190 species will require conservation of a large expanse of Basin forest. Undoubtedly this conclusion also applies for other taxa that have not been surveyed as thoroughly as birds.

THE PEOPLE OF THE LAKEKAMU BASIN

There are four major landowner groups residing within the Lakekamu Basin: the Biaru; the Kamea; the Kovio; and the Kurija. All groups currently utilize resources within the Basin but the boundaries of their land claims are not agreed upon and are often strongly contested. The technical report provides extensive background details on the complicated history of each group in the Basin, but does not provide information that could be used in legal settlement of any ownership disputes in a court of law.

Shifting agriculture is the main form of subsistence, but its ecological impact is limited due to the relatively small population size and the fact that gardens are largely confined to river edges leaving forest interiors uncut. Forest products are important for many uses including food, shelter, medicine and ceremonial purposes. Cash revenue comes primarily from sale of betelnut outside the Basin, some small-scale gold extraction and the sale of a small amount of game and animal products sold. Visiting researchers and ecotourists are providing a growing source of income through employment and sale of garden crops.

Large-scale development in the form of timber, gold-mining or plantation agriculture has been considered at various times. However, the different landowner groups often disagree on the details of such development and indeed there is often lack of consensus within a group on the desirability of large-scale development due to the recognized importance of forest resources and clean water.

OVERALL SUMMARY

The Lakekamu Basin encompasses roughly 2500 km² of pristine wilderness lowland rain forest. The forest is sparsely inhabited, less than 1 person per km². The impacts from human use have been minimal and are probably sustainable at current levels except in regions of heaviest use. Populations of game species, such as cassowaries, seem vigorous. The rivers are unpolluted, and essentially devoid of invasive exotic fish. Human consumption of aquatic resources, such as fish and crocodiles, is also still at a relatively low and potentially sustainable level (immediate study is highly recommended). Thus, the Basin represents one of the best and largest tracts of lowland forest in pristine condition in PNG. The size and quality of the forest and the low number of people who live within it make it an outstanding conservation opportunity.

The biodiversity within the Basin is globally significant. The Basin is exceptional in terms of numbers of species (*e.g.*, the richest ant fauna in the world), and it is outstanding in terms of endemism (New Guinea's biota is essentially unique except for a few species shared with Australia's threatened fragments of remaining rainforest). However, what is most significant in terms of conservation is the size and condition of the rainforest habitat. Within the Lakekamu Basin and adjoining hills lies a vast, intact ecosystem. Conservationists, land managers and ecologists around the world agree that successful conservation must proceed at the ecosystem scale. The crocodiles and large fish of the lower Lakekamu Basin depend on pure water resources from the upper Basin. Birds like parrots and raptors forage over huge home ranges within the Basin. Many rainforest plants exist in very low densities, requiring many square kilometers just to sustain a viable population. To maintain a complete array of existing species, much of the large Lakekamu Basin ecosystem must remain relatively intact.

The impact of the resident human population at present is not excessive; the population is small but growing, gardens are largely confined to river edges, and conflicts between landowner groups prevent permanent settlement in some areas. Unlike many globally significant tracts of rainforest, there is not impending external threat of destruction. There are no roads through the Basin, making logging or plantation agriculture more difficult, but access is fairly easy via river or plane from Wau or Kerema—a plus for considerations of ecotourism or extraction of light non-timber forest products. The timber of the Basin is not particularly valuable or easily extracted; logging companies might clear other parts of the country first. Nonetheless, given the ever-growing demand for timber and the ever-shrinking sources for timber globally, it is inevitable that loggers will soon eye the Lakekamu Basin. It would serve conservation well to have viable economic alternatives to logging in place and producing *before* logging becomes an issue. There is some gold in the Basin and large-scale extraction would seriously impact water quality. To date, only small-scale, low-impact extraction has occurred. The situation bears close monitoring, but it is possible that landowner disputes could stall any larger gold projects in the immediate future.

The 1992 DEC Conservation Needs Assessment listed the Basin as a priority area in terms of biodiversity and an as area where the lack of scientific knowledge is serious. Since then, the Lakekamu Conservation Initiative, administered by FSP-PNG and CI, has made progress in the realm of conservation and the RAP survey has begun the immense task of addressing our lack of scientific knowledge. The improving landowner relations in the Basin, the development of permanent research infrastructure, and the completion of a major biotic survey all bring the Lakekamu Conservation Initiative to a new threshold.

Research and training in the Lakekamu Basin

can and should now become a major priority. The RAP survey, combined with previous work in the Basin, lays a solid foundation for expanded research efforts. Initiation of long-term research at the Ivimka Research Station would provide a steady source of income that goes directly to and is equitably spread among Basin landowners. Research does not provide an unmanageable surge in income that can be damaging to small village societies and economies, nor does it damage the environment. Research provides an ideal vehicle for training PNG's young biologists (as has already begun with several UPNG post-graduates at the IRS) and provides a medium where visiting scientists can mentor PNG students. Lastly, scientific data is sorely needed to answer pressing conservation issues across PNG. The discovery of 35-47 new species and several new genera in just one month at one site is a potent indicator of the magnitude of our ignorance. Research in the Lakekamu Basin could become a vital resource for correcting our ignorance and provide essential data for guiding PNG's development.

A series of conservation recommendations follow that are derived from specific findings of the RAP scientists. There are three essential, all-encompassing recommendations:

1) Conservation requires a broad approach across the entire Basin ecosystem. Making the Lakekamu Basin the globally outstanding conservation site it can be requires appropriate conservation activities over the whole Basin. Some species require extensive forest for their survival, others rely upon ecological interaction between regions within the Basin for maintenance.

2) Integrate, equitably, all the people of the Basin into conservation and economic endeavors that support and profit all parties involved. Successful conservation in PNG is by necessity landowner-based. The situation in the Basin wherein there are four major landowner groups with conflicting land ownership claims means that no party can perceive itself as excluded from the process, or contentious issues will be aggravated. Active participation and tangible benefit by the residents will ensure the long-term viability of the conservation initiative and ultimately the Basin ecosystem.

3) Develop the research activities at the Ivimka Research Station. Research leads the way for other conservation-oriented activities, (*e.g.* ecotourism, non-timber forest products) and is an excellent means of establishing long-term bonds between landowners and conservationists. Research has spear-headed many successful conservation efforts worldwide (*e.g.*, Manu National Park in Peru, or the Danum Valley in Malaysia). With adequate external funding, the Lakekamu Basin could become a globally significant rainforest conservation and research center.

RECOMMENDED CONSERVATION AND RESEARCH ACTIVITIES

The following list of recommendations is drawn from the text. Most recommendations are composites of suggestions made in several chapters and listed roughly as they appear in the text. Recommendations are not listed in order of priority.

• Continue and improve collection of meteorological data at the IRS.
• Detailed ecological studies are needed for virtually all taxa. However, species that are consumed by people, threatened, widely-ranging in the Basin, or naturally occur at low density are particularly worthy of study to meet conservation needs.
• Successful conservation of the biota within the Basin will necessitate preservation of a large area, encompassing the full array of habitat mosaics within the Basin and preferably including adjacent foothill regions.
• Strive to avoid development projects that entail large-scale forest conversion. Due to the patchy distribution of species, such projects could cause local extinction and facilitate invasion of weedy and exotic species.
• Strive to prevent the introduction and establishment of invasive exotic species (ants, fish, weeds, *etc.*) in the Basin.

• Utilize the IRS infrastructure and study area as a training location for young biologists in PNG. Training is strongly needed in basic taxonomy because PNG needs scientists capable of finding and identifying its native biota.

• Encourage scientists visiting from overseas to train/mentor students. The IRS is especially well-suited for combining field research and mentoring.

• The IRS study area is well-suited for research into the role of natural disturbance in the maintenance of high species diversity.

• Priority should be given to a thorough botanical inventory of the Basin, prior to undertaking other vegetation or ecological studies. Such an inventory could provide an excellent vehicle for botanist training in PNG.

• Botanical surveys are more strongly needed in the Papuan (southern) part of PNG than the Mamose (northern) part of the country.

• Incorporate aquatic organisms, like fish and odonates, in any monitoring of water quality that will be undertaken in the Basin.

• Establish long-term monitoring of amphibian populations as part of the study of global declines in amphibians.

• Building on the solid foundation begun by the RAP, a long-term avian monitoring program using mist nets and point counts should be initiated at the Ivimka Research Station.

• Work with local landowners and hunters to develop a hunting moratorium over a broad area surrounding the Ivimka Research Station.

• Monitor use and trade of non-timber forest products by the residents of the Basin.

• Review all pertinent records from the colonial era to assess the history of occupation in the Basin by the different landowner groups.

• Assess the long-term history of settlement in the Basin with archaeological survey work and excavation.

• Conservationists must take into account the complex social history of the Basin and develop new forums for discussion and planning that transcend traditional social boundaries and obstacles.

INTRODUCTION

Conservation International's Rapid Assessment Program (RAP) evolved in Latin America to meet the critical need for quick identification of priority areas that could then be targeted for conservation action. Fifteen RAP surveys in Latin America (and Madagascar) have helped gazette six national parks or protected areas in five countries. In PNG, there also is a strong need for such assessments, especially in light of the rapid expansion of foreign timber companies. However, due to the strong land tenure laws of PNG, no matter how extraordinary an assessment, PNG cannot gazette a national park. Conservation in PNG requires direct, enthusiastic, and long-term participation between landowners and conservationists. Thus, the Lakekamu survey was an adaptation of the RAP to the special needs of PNG.

This RAP was not intended to identify or justify a conservation priority area; Lakekamu was already an established priority (Beehler 1993) and the Lakekamu Basin Conservation Initiative had been in place for three years when we initiated the RAP. Instead, this RAP was designed to provide information that would be useful for the development and maintenance of the existing conservation initiative. In the next few decades, the world's only remaining wild and intact ecosystems will be largely confined within the boundaries of conservation areas. As this occurs, the role of RAP and conservation biologists increasingly turns to providing data for the management of conservation areas and less toward exploration of truly unknown places. Our hearts go out to biologists of the future who will never know, as we have, the excitement of exploring a rainforest where no biologist has trod.

THE LAKEKAMU CONSERVATION INITIATIVE

The aim of FSP-PNG's Integrated Conservation and Development Program (ICAD) is to promote development and conservation by establishing direct linkages between an increase in the socio-economic well-being of landowners and conservation of natural resources. The FSP-PNG program focuses on the Lakekamu-Kunimaipa Basin in Gulf Province, funded largely by USAID through the Biodiversity Conservation Network. The program integrates conservation and development, creating economic enterprise such as eco-tourism and the marketing of non-timber forest products. It promotes the conservation of the lowland forest of the Basin and its upland watershed by building the capacity of the local landowners to be leaders in their own development and in conservation of the resources through training and on-site assistance. Conservation International and the Wau Ecology Institute collaborate closely with FSP-PNG on the project.

Three main activities have been FSP-PNG's focus:

1) the development of science/research tourism
2) development of adventure (eco-) tourism
3) providing education and training for the local communities.

C. G. BURG

B. BEEHLER

View looking upstream on the Sapoi River near the Ivimka Research Station. The edges of the larger rivers like this are home to second growth and successional species that are rare in the forest interior.

Typical appearance of alluvial forest understory in the Lakekamu Basin. Note the presence of buttressed trees and woody lianas, hallmarks of many tropical forests worldwide.

G. ALLEN

The Clinging Goby (Lentipes watsoni), a species first discovered and described from material collected on the RAP. This is the first member of the subfamily Sicydiinae to be recorded from southern New Guinea.

G. ALLEN

G. ALLEN

The Ivimka Creek where it leaves the foothills. Generally the forest canopy closes above such small streams, creating a different aquatic and edge environment than found on the larger rivers. Small streams like this meander through the alluvial forest, increasing forest dynamics by undercutting trees as they change course.

Aerial view of the upper Lakekamu Basin around the Ivimka Research Station (metal roof visible). Above the station is the undulating terrain where the hill forest occurs and below the station is the alluvial forest. Differences in the canopies of these (and other) forest types in the Basin can be noted from the air. On the left side of the image you can see the near junction of the Avi Avi (leftmost) and Sapoi Rivers.

C. G. BURG

A female-plumaged Twelve-wired Bird of Paradise (*Seleucidis melanoleuca*). This species is commonest in the lower parts of the Basin in regions that flood.

THE IVIMKA RESEARCH STATION

Construction of the Ivimka Research Station (IRS) began in March 1996. By mid-October 1996, when the RAP began, the building was nearing completion and work continued as the biologists conducted their field studies. At the time of publication, the station is essentially complete. There are three goals for building the station and developing research in the area.

1) The IRS can provide a modest, but steady source of income for the local landowners based on non-destructive use of their forest—research and the extraction of data.

2) The IRS can act as a base for field-training of PNG biology students.

3) The IRS can facilitate new field research in an intact lowland rainforest ecosystem and promote long-term field studies. A comfortable, well-equipped field station will promote these three objectives. A long-term commitment to conserving the area by the landowners would enhance the research potential at the site.

Details on the IRS facilities and arrangements for visiting or conducting research there can be obtained by contacting FSP/PNG.

GOALS OF THE LAKEKAMU RAP

There were six principal goals for choosing the Lakekamu Basin as the site of the RAP survey:

1) Produce data that would serve as a solid foundation and stimulus for future research at the IRS by attracting more researchers to the IRS and making their work easier.

2) Verify the importance of the Lakekamu Basin in terms biodiversity, both on a regional (PNG) and global geographic scale.

3) Produce data that would assist the local people and their partners with their conservation goals.

4) Establish baseline data that can be used for future teaching, monitoring and evaluation.

5) Improve our knowledge of the rainforest biota of PNG in general.

6) Improve the research infrastructure

(*e.g.*, build equipment inventory, map trails, establish permanent vegetation plots, *etc.*) of the IRS. The sixth goal was accomplished during the RAP survey itself. It is hoped that this report and the modest collections produced by the RAP survey will help meet the first five goals.

THE STUDY AREA

The RAP was undertaken 15 October to 15 November, 1996 in the area surrounding the Ivimka Research Station (146° 29' 45"E, 7° 44' 05"S) in Gulf Province, near the boundary with Morobe Province, 10 kilometers SSE of Tekadu Airstrip. Additional survey data were collected during the UPNG training course that occurred at the same site 16 November to 12 December, 1996.

Most survey work was done along the trail system of the IRS study area (See map page 22). The trail network involves two pre-existing trails and several new trails for researchers. The main foot trail connecting Tekadu to Kakoro passes just below the IRS, passing through the study area and is frequently traveled by the local residents. This track diverges from the old Bulldog Road by the base camp occupied by staff and visitors prior to the completion of the IRS. The Bulldog Road continues south along the Sapoi River (= East Branch of the Avi Avi River on some maps), eventually bending westward to Bulldog and the abandoned airstrip where the Sapoi, Avi Avi and Tiveri Rivers meet (See map page 22). Six working transects were cut running due east of the Bulldog Road one kilometer apart starting 1 km south of the former base camp. Each transect was one km long, but the third was extended a second kilometer to the east. A trail to the SSW was cut off from the 3 km mark on the Bulldog Road to access the different forest growing between the road and the Sapoi River at this point. Around the old base camp, there is an inner loop trail and an outer loop trail that connect the Kakoro track to the Bulldog Road. Another loop (the garden loop), encircles an abandoned garden on the north side of the Kakoro track. The ridge trail joins the Kakoro track near the old base camp, ascends to the IRS and continues upridge, crossing the headwaters of the Ivimka Creek, to about 400 m elevation where trail cutting ended. This trail could be extended to higher elevations into the foothills of Mount Lawson. The Ivimka Creek forms in the foothills above the IRS then meanders through the level Basin forest, crossing the Kakoro trail several times, before it joins the Sapoi River.

The RAP survey mostly confined itself to areas within 3 km on the Bulldog Road and to the end of the ridge trail (See map page 22). When cleared, the level, dry grade of the Bulldog Road permitted fast access to the transects south of the IRS. These transects penetrate forest that is infrequently visited by the local inhabitants, as it is between the Kakoro track and the Avi Avi River used as the track to Nukeva and Okovai. Overflights of the area indicated that this area is typical of the upper Lakekamu Basin. Further south in the Basin there are large tracts of swamp forest that are probably inundated much of the year and probably contain lower tree species diversity. In the upper Basin drainage is better, with numerous meandering small streams draining the forest and few areas of frequent inundation. These meandering streams, even the small ones under closed canopies, contribute to forest dynamics and seral heterogeneity by eroding tree bases and increasing treefall rates above that caused by windthrows and senescence. Additionally, there are seral gradients along the larger river edges and the 50-year old regrowth along the Bulldog Road.

GEOLOGY

There are predominantly four major geologic formations in the Basin and nearby surrounding hills (Dow *et al.* 1974). The Basin itself is comprised of Quaternary alluvium from the Holocene: unconsolidated fluvial sediments, gravel and sand. The hills just surrounding the IRS to the north and west are a small intrusion of Edie Porphyry from the Pliocene. These are comprised of biotite and hornblende andesite and dacite, porphyry stocks and dykes. Opposite the IRS to the west, across the Avi Avi River is Morobe Granodiorile (as is Mount Lawson) of Miocene origin. This is comprised of granodiorite, adamellite, subordinate monzonite, diorite and pegmatite. The majority of the hills to the north and west of the Basin, including the Chapman Range are Owen Stanley Metamorphics. These are predominantly comprised of low-grade metamorphics: slate, phyllite, quartz-sericite schist.

CLIMATE

Temperatures in the Basin range from 20° to 29.4° C (Oatham and Beehler 1997). Yearly rainfall in the area has been reported as about 3.5 m per year (Beehler 1995), but incomplete rainfall data collected by LCI project staff in 1996 and early 1997 indicated an annual rainfall over 5 m. Rainfall during this period was probably higher than normal (C. Makamet, personal communication). Judging by the vegetation on of the Basin, it is likely that rainfall does not normally exceed 5 m per year. Seasonality is not strong in the Basin and the limited rainfall data support this. Unfortunately, the two months, June and July, for which data are absent are thought to typically be drier. Clearly more, and thorough meteorological data need to be collected. A good foundation is begun by the data collected by the FSP-PNG project staff. We recommend that the LCI and IRS staff continue and expand their collection of meteorological data. Accurate and rigorously-collected climatological data will substantially improve the quality and scope of research capabilities at the IRS.

HISTORY AND THE
BULLDOG ROAD

EXPLORATION BEFORE THE WAR

When the Australians took control of German
New Guinea at the onset of World War I, a
German miner, Hellmuth Baum, refused to sur-
render and disappeared into the rugged mountains
around Wau. He explored and prospected in the
area for more than 15 years before being killed by
Kukukukus (see Appendix 19). The Australians
sent patrols into Kukukuku country in 1931 and
1932 to try to bring Baum's murderers to justice.
These patrols established a trail from Wau to
Kudjeru and mapped a track to the Lakekamu
River. As more prospectors wished to enter
Uncontrolled Territory, the District officers
increased their activities in the area in order to
make it safe for exploration. However, viable
gold deposits were not found and outside contact
for the native inhabitants was limited to a few
prospectors and the occasional kiap patrol.

THE WAR

The first air attacks on mainland New Guinea
were in January 1942, causing the evacuation of
Lae and Salamua; with many people fleeing to
Wau and Bulolo. Japanese land forces put ashore
at Salamua and Lae in March and met little resistance. The Allies suspected Japan would march
against Port Moresby, but were unsure from
where the attack would come. The first planned
invasion of Port Moresby was to be by sea, but

Japan's heavy losses in the Battle of the Coral
Sea in May 1942 caused postponement of the
invasion and an overland attack was chosen for
the next attempt.

In the spring of 1942, the Japanese had uncon-
tested control over northern New Guinea, the
Bismarcks and Solomons. The Australians in the
south were only separated from the Imperial
Army by the steep Central and Owen Stanley
Ranges. If they were to eventually remove the
Japanese from the north they would have to either
advance over the mountains or around the southeast peninsula. The Japanese faced the same
dilemma for clearing the Australians from Port
Moresby to open Australia to invasion. In May,
the Allies airlifted troops to Wau-Bulolo in order
to attack the Japanese in the lowlands, but strong
counterstrikes by the Japanese stopped any further attacks out of Wau.

The Japanese launched a second move on
south New Guinea in July, moving over the Owen
Stanleys via the Kakoda Trail. Fierce fighting
occurred and the outnumbered and under-trained
Australian forces were pushed back almost to
Port Moresby. However, the long supply-line and
poor conditions made fighting almost impossible
for the Japanese and Port Moresby was held. The
Allied commanders learned how difficult it would
be to move over the mountains to re-take the north.

The Allies moved around the southeast peninsula via Milne Bay and eventually prevailed after
very heavy fighting at Buna, opening the way
for recapturing Lae. In the spring of 1943 the

Japanese planned to counterstrike the Allies via Wau, and made it as far as the Wau airstrip, but the Allies managed an intensive airlift to support Wau and fought back the Japanese, who had a difficult overland supply-line to the coast. Another attack on Wau was planned, the last hope of the Japanese to regain the ground they had lost in New Guinea. However, many of the incoming reinforcements were killed and the rest forced to retreat to Rabaul in the Battle of the Bismarck Sea. By the summer of 1943 the Allies had taken Salamua and Lae and Japanese were in retreat in the southwest Pacific.

World War II brought many fundamental changes to the island of New Guinea; tens of thousands of American and Australian soldiers passed through areas that had rarely been visited by outsiders. The construction of the Bulldog Road brought many Australians into the Lakekamu Basin, a region that had only sporadically been visited by Australian kiaps and miners, if visited at all.

THE BULLDOG ROAD

Civilians fled the attack of Lae up the Bulolo Valley to Wau. Here they sought a way to the south coast. In Wau they met with one of the kiaps who had helped explore the region searching for the miner, Baum's, murderers in Kukukuku country. With the help of three surveyors who knew the area they fled over the mountains to Kudjeru and down to the lowlands, connecting to the coast by boat on the Lakekamu River.

The successful movement over the mountains to Moresby of 250 civilians (considered unfit for military service) caught the attention of the military command. In March 1942 orders were made to survey a road following the path they used. The Allies needed some way to maintain their garrison in Wau and needed options for the anticipated invasion to recapture the north. In April a line of communication, a series of camps, was established between Bulldog and Wau. In August serious plans began for construction of a jeep track from Bulldog to Wau. This road could

become an important strategic asset to rout the Imperial Army from the north.

Road construction began in January 1943. The route was changed to go via Edie Creek rather than via Kudjeru (the old Bulldog walking track), partly due to engineering considerations and partly due to the Japanese movement into the Kudjeru-Wau stretch during their spring offensive on Wau.

The road started at Bulldog at about 35 m elevation, passed through Ecclestone Saddle at 2885 m and ended at Edie Creek, about 100 km from Bulldog. The road was completed in just eight months. Many PNG nationals worked as laborers and carriers on the project, with more than 500 carriers and 1000 laborers during peak construction. Many of the work force were from the lowlands and suffered illnesses (e.g., bronchitis and pneumonia) when working in the cold, damp conditions at higher elevations. Some tensions developed, allegedly over the abuse of a local woman by soldiers, leading to the death of one Kukukuku; later two soldiers were killed near the RAP study area. A local informer told Dr. Mack the story of how a soldier who raped a local woman was killed, his body was hidden and never found. The Japanese bombed and strafed Bulldog during construction, killing one national and causing the bulk of the work force to abandon the project. Due to the haste at which the road was built, it could not be constructed in a way that would minimize maintenance or ensure its viability.

The road was used as a supply route to Wau for only six months. With the hard-won victories at Buna and Salamua on the north coast and the Allies' growing strength at sea, it became more viable to move men and materials by ship rather than the arduous route over the Bulldog Road. The road was abandoned in February 1944 never having served any significant tactical purpose. After the war it was noted that the same amount of cargo as the road transported could have been moved with just 25 C-4 air transport planes.

The remnants of the Bulldog track are a conspicuous feature of the upper Lakekamu Basin and the field study area around the Ivimka Research Station. The road bed itself is a major trail in the study area and connects the station to the airstrip at Tekadu. The forest was cleared 33 feet on each side of the road center. Twenty feet to each side of center was cleared of stumps. The regrowth of vegetation along the roadbed is impressive. Fifty years after being abandoned, it requires a skilled eye to habitat nuances to discern what is regrowth and what is primary forest. It should be noted that substantial portions of the road through the alluvial forest were underlain with corduroy. Presumably, this derived from trees cut for the road path, but possibly additional tree-felling occurred beyond the 66 foot road swath for splitting corduroy. The drainage ditches and road grade now form pools in places that are well-suited as frog breeding sites. There are still numerous relics of the war in the area, old steel beams that used to be bridges, crankcases, chains, engine blocks and other mechanical debris can still be found, often in surprisingly good condition after 50 years. Another possible reminder of the war are the presence of exotic species, such as the ant *Cardiocondyla nuda,* that might have been inadvertently transported to the area with war supplies.

Beyond it's historical interest, the Bulldog track presents both an opportunity and a threat. The roadbed increases the utility of the Ivimka Research Station. It provides easy and quick access to an extensive stretch of forest and offers the opportunity for scientists to study regrowth of an abandoned road. Because the track is not used by the local people south of the IRS, the wildlife along it has not been scared away or hunted (as has been the case along the Tekadu-Kakoro trail). The Bulldog Road also could be promoted as a bush-walking track, attracting small numbers of hardy eco-tourists to this remote part of the country. The threat posed by the road would come from any timber companies that view the old roadbed as an opportunity to facilitate timber extraction.

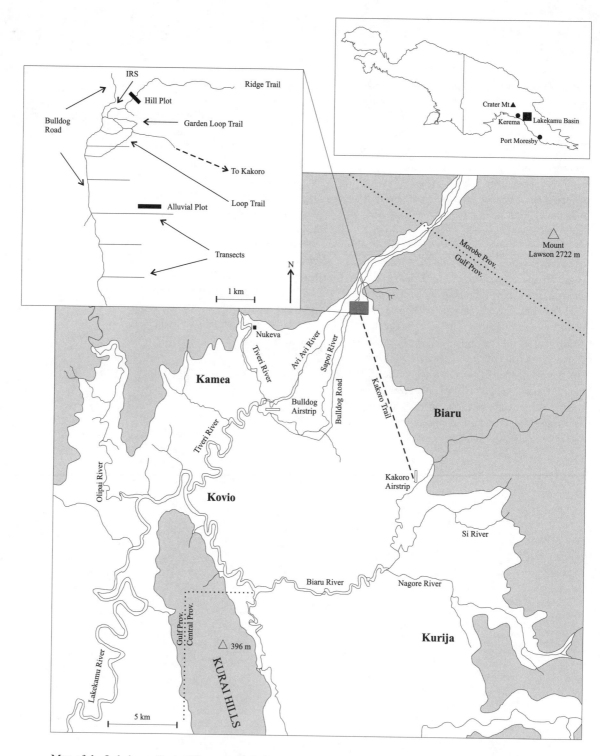

Map of the Lakekamu Basin. The upper right inset shows the position of the Lakekamu Basin in PNG, the mouth of the Lakekamu River is just below Kerema. The large map shows detail of the upper basin. The general areas occupied by the four major language groups in the basin are indicated by the names of those groups in boldface. The shaded areas represent elevated ground above the 200 m contour. The upper left inset shows details of trails in the study area. Position of the Ivimka Research Station (IRS) is indicated with an arrow. Transects were made at one kilometer intervals along the Bulldog Road. Positions of the two vegetation plots are shown. Mist nets for birds and bats and mammal traps were set along the three northernmost transects and the ridge trail.

CONSERVATION INTERNATIONAL

Rapid Assessment Program

TECHNICAL REPORTS

VEGETATION PART 1: A COMPARISON OF TWO ONE-HECTARE TREE PLOTS IN THE LAKEKAMU BASIN

(J. Alexandra Reich)

Chapter Summary

• We studied forest structure, species diversity and soil characteristics in two one-hectare plots in the IRS study area, one in hill forest and the other in alluvial forest. Plots were 20 m x 500 m with 25 subplots measuring 20 m x 20 m.

• We marked and measured all trees ≥10 cm DBH, and collected voucher specimens from all marked trees for species identification. Lianas were excluded from the study.

• Roughly 253 species were censused: 574 individuals of 182 species on the alluvial plot, and 759 individuals of 156 species on the hill plot. Stem density on the hill plot was significantly higher than on the alluvial plot. Stem size class distributions between the two plots did not differ.

• Most species were "rare." In the alluvial plot, 87.4% of the species had 5 or fewer representatives, and only 1.1% (2 species) had more than 20 individuals. Similarly, 75% of the hill species were represented by 5 or fewer individuals, and 4.5% (7 species) have more than 20 individuals.

• Soil characteristics were essentially typical of tropical rainforests. Soil nutrient contents, pH (4.0 for both), and cation exchange capacities (alluvial = 5.8 and hill = 6.8) were all low; iron and aluminum were high.

• Compared to three other plots in the Basin (Oatham and Beehler 1997), the Ivimka plots had more stems but fewer large trees.

• Through this study we have added to the list of rare species known to occur in the Basin and further demonstrated the heterogeneity within the Basin. The high diversity, rarity and heterogeneity of plant taxa within the Basin testify to the need for conservation. Successful conservation will require broad coverage to encompass the variety of vegetation types occurring within the Basin.

Introduction

In the Asian tropics, PNG is the seventh largest country (out of 19), and holds roughly 450,000 km² (12-15%) of the moist and lowland tropical rainforests found in the region; making it second only to Indonesia (Davis 1995a). Considering New Guinea as a biogeographic unit (combining PNG and Irian Jaya) elevates the region to greater prominence: over 800,000 km² of forest or roughly 45% of all forest in Malesia (the biogeographic area extending from Malaysia to New Guinea). Thus New Guinea contains more forest than the rest of Indonesia, Malaysia, or Indochina. It is one of the few remaining, large tropical wilderness areas.

Because of its geologic history, New Guinea has a unique flora, combining taxa from the Oriental and Australian biogeographic regions. At least 25,000 species of vascular plants occur in

New Guinea, the vast majority of which, 20,000 or more, occur in PNG (R. Johns, personal communication, Höft 1992). Thus, roughly half the plant species of the Malesian region are found here. This translates to an astounding figure: roughly 10% of the world's plant species occur in New Guinea.

Among these, most species are found nowhere else in the world. There are 93 endemic genera (Osborne 1995) in PNG and 70-80% of the plant species are thought to be endemic (R. Johns in Davis 1995b). However, these estimates are based on a small number of studies, and a paucity of specimen collections. It is likely that expanded survey work will reveal many new species from PNG and Irian Jaya. Quite likely many of the taxa considered endemic to PNG or Irian Jaya might be found in the other country with adequate survey work. Nonetheless, the global importance of PNG's floral diversity will only increase with further study. Many taxa are shared with Australia's small fragments of rainforests. As these pockets of rainforest degrade, due to the expected consequences of fragmentation (Lovejoy et al. 1986), the large intact ecosystems of New Guinea become that much more important.

Few studies have examined the structure and diversity of PNG forests (e.g. Paijmans 1970, Johns 1986, Wright et al. 1997), and fewer studies have concentrated on the lowland rainforests (Oatham and Beehler 1997, Kiapranis 1991). Nonetheless, it is clear that the rainforests of PNG are similar to other rainforests in that they are very species rich, with many species naturally occurring in low densities (see Discussion). Rainforest trees are vital to many people of PNG, providing many essentials, from food and medicines to construction and fuelwood. Because logging companies are securing timber rights over large areas and forests are being harvested at an alarming rate (Nadarajah 1993), it is urgent that more information be obtained on the structure and ecological functions of PNG's native forests. Such information is a prerequisite for prudent management of forest resources.

Through sampling two one-hectare plots, we intended to investigate forest structure and tree diversity in a portion of the Lakekamu Basin near the Ivimka Research Station. These data will assist future research and conservation activity around the IRS. Information from this study augments previous work in other parts of the Lakekamu Basin (Oatham and Beehler 1997). As more vegetation studies are conducted in the Basin, it will become easier to develop the management priorities and goals for the Lakekamu Conservation Initiative and to manage other rainforest sites in PNG. Finally, the data from this study will help improve our knowledge of the richness of PNG's lowland forests relative to other parts of Asia and the Neotropics that are better-known. Studying rainforests in PNG enhances our knowledge of rainforests worldwide, aiding conservation efforts within and outside of PNG (Oatham and Beehler 1997, Soepadmo 1995).

Methods

We randomly selected two sites from previously existing trails (the Bulldog Road and the Ridge Trail, See map page 22) for a one-hectare plot in the alluvial and hill forests, respectively. The alluvial plot is 3 km south, and the hill plot is approximately 0.5 km northeast of the Ivimka Research Station. We delineated belt transects measuring 20 m x 500 m, consisting of 25 subplots (20 m x 20 m) for intra-plot comparisons. Using an altimeter, we recorded the elevation of each subplot, and with a compass/ inclinometer, we measured slope and aspect. From the center of each subplot, we subjectively estimated canopy cover. We permanently marked and measured all trees $\geq$ 10 cm diameter at breast height (DBH, breast height = 1.4 m above ground). For those trees with large buttresses, we measured diameters just above the buttresses, and recorded the height at which the measurement was taken. In the case of *Pandanus* spp. with multiple stilt roots at 1.4 m high, we measured the diameter of all living prop roots, calculated their total basal area and calculated a measure of diameter from the total basal area (Wright et al. 1996).

We collected voucher specimens from all marked trees for identification and storage at the Forestry Research Institute (FRI) in Lae, PNG. In the event of multiple individuals of a positively identified species, we collected one representative specimen for that species. We pressed and stored

Thus New Guinea contains more forest than the rest of Indonesia, Malaysia, or Indochina. It is one of the few remaining, large tropical wilderness areas.

all specimens in 75% ethanol until transported to a suitable drying facility.

Kippy Demas identified dried specimens to genus or species at FRI. Where identification to species level was not possible, and where sufficient differences were apparent, congeners were divided into morpho-species. We measured species diversity for both plots using the Shannon Diversity Index (Zar 1984). As a measure of the relative contribution of each family to the plots, we calculated Family Importance Values (FIV, Mori *et al*. 1983), an index that incorporates relative diversity, relative density, and relative dominance. Species *vs*. area curves were derived from the cumulative number of species in each 20 m x 20 m subplot.

We collected, dried, and sieved soil samples at 0 cm and 20 cm depths from one random point in each subplot. Soil samples were analyzed for nutrient and mineral content, pH, salinity, and cation exchange capacity by Micro-Macro International, Inc. (Athens, GA). Samples from the 20 cm depth were treated for exchangeable macroelements with 0.5N ammonium acetate and with 0.0005N DTPA for trace metals (Page *et al*. 1982). Extracts were then analyzed with an inductively coupled plasma spectrometer. Nitrogen content was determined with the microkjeldahl technique (Hesse 1972).

We calculated mean tree density per subplot, and tested for a difference between the two plots with a student's t-test. We plotted size class distributions in 5 cm increments, and using the nonparametric Kolmogorov-Smirnov Two-Sample Test, we tested the differences in the size class distributions of the two plots (Sokal and Rohlf, 1995). Where applicable, we made comparisons between our data and published data for PNG, and other tropical sites. Statistical tests were performed with Sigmastat 1.01.

Results

Density and Dominance

We found 575 trees (DBH $\geq$ 10 cm) in the alluvial plot and 759 trees on the hill plot (Table 1). Stem density (mean number of trees per subplot) was greater on the hill plot than the alluvial plot (Table 1; t= -4.20, df=48, P=0.0001). However, basal areas (dominance) did not differ substantially between the two plots. Total basal area of trees was 28.46 m^2 in the alluvial plot, and 32.04 m^2 in the hill plot (Table 1). Size class distributions between plots did not differ (Figure 1; KS = 0.062, KS^{crit} = 0.075, n^1 = 575, n^2 = 759, P $\geq$ 0.05). Neither plot had as many large trees as on three other plots in the Lakekamu Basin (Table 1). The stem density and basal area of the Ivimka plots are within the range of what is typical for tropical rainforests around the world (Table 2).

Table 1. Summary statistics for diversity, density and size of woody stems on vegetation plots in the Lakekamu Basin. Alluvial and hill plots from this study at Ivimka, Si and Nagore plots from Oatham and Beehler (1997).

PLOT	SHANNON INDEX **	# OF SPECIES	# OF FAMILIES	# TREES/SUBPLOT (standard deviation)	# TREES $\geq$ 10 cm DBH	MEAN DBH (cm)	# TREES > 60 cm DBH	# TREES > 100 cm DBH	BASAL AREA (m²/ha)
Alluvial	4.86	182	52	23.0 (5.62) ***	575	21.1	10	1	28.46
Hill	1.90	156	49	30.4 (6.73) ***	759	19.9	13	0	32.04
Si River*	1.75	93	34	n.a.	392	28.6	43	20	49.32
Nagore N*	1.96	149	42	n.a.	426	23.3	26	1	37.67
Nagore S*	2.01	178	42	n.a.	482	22.1	23	2	28.72

n.a. = not available
* Note: the Oatham and Beehler data include lianas whereas this study did not.
** Note: Shannon Index calculated using Log^{10}.
*** There is a significant difference between the mean number of trees per subplot (t = -4.20, df=48, P=.0001)

Table 2. Tree densities, basal areas, species densities, regions and forest types from studies in comparable tropical forests throughout the world.

REGION	FOREST TYPE	DENSITY #TREES/ha	BASAL AREA (m²)	DIVERSITY #SPECIES/ha	CITATION
PNG - s	Alluvial/Lakekamu	575	28.5	182	This study
PNG - s	Hill/Lakekamu	759	32.0	156	This study
PNG - s	Lowland/Lakekamu	433	38.6	178	Oatham and Beehler 1997
PNG - n	Low montane	575	29.2	166	Paijmans 1970
PNG - c	Mid montane	693	37.1	228	Wright et al. 1997
Asia	Low/dipterocarp	375	n.a.	134	Poore 1968
Asia	Alluvial	615	28	225	Proctor et al. 1982
Asia	Heath	708	43	110	Proctor et al. 1982
India	Low montane	635	39.7	91	Pascal and Pelissier 1996
Americas	Alluvial	580	n.a.	300	Gentry 1998
Americas	Alluvial	650	n.a.	204	Gentry 1998
Americas	Alluvial	526	33.5	165	Gentry 1998
Americas	Lowland	446	n.a.	269	Lieberman 1984
Americas	Mid elevation	891	n.a.	200	Mori et al. 1983
Americas	Upland	575	29.1	172	Phillips et al. 1984
Americas	Terra firma	432	n.a.	179	Pires et al. 1953

n = northern, c = central, s = southern, n.a. = not available

Conservation in the Basin that includes adjoining hill forest could preserve a vastly greater range of diversity in a relatively narrow fringe of surrounding hill forest.

Diversity

There were 253 species or morpho-species identified on the plots: 182 species in 104 genera and 52 families in the alluvial plot, and 156 species in 92 genera and 49 families in the hill plot. (Appendix 2). The continued positive slopes in the species-area curves suggest that one hectare is not sufficient for a thorough sampling of the local tree community (Appendix 1A). The Shannon Diversity Index for species in the alluvial plot is 4.86 and for the hill plot is 1.90. There were 86 species (47.25%) in the alluvial plot represented by only one individual, and 64 species (41.03%) in the hill plot (Figure 2). Further, 87.36% of the alluvial species, and 75% of the hill species, have 5 or fewer individuals. Only 2 species (1.1%) in the alluvial plot and 7 species (4.5%) in the hill plot have ≥ 20 individuals (Table 3). Only one species, *Parastemon versteeghii* (Chrysobalanaceae) is among the ten most common species in both plots, with 15 individuals in both plots (Table 3).

Seven families were shared among the ten highest Family Importance Values for both plots (Table 4). Only three of the top ten families at Ivimka are also in the top ten at three other sites in the Lakekamu Basin (Table 5). Eight of 21 dominant plant families are in the top ten of only one of the five inventoried sites in the Lakekamu Basin (Table 5) yet one of these, Datiscaceae on the Si plot, has the fifth highest of fifty top FIV scores in the Basin.

Table 3. Number of individuals for the ten most common species in the alluvial and hill plots.

FAMILY	ALLUVIAL PLOT SPECIES	NUMBER INDIVIDUALS	FAMILY	HILL PLOT SPECIES	NUMBER INDIVIDUALS
Sapotaceae	*Pouteria firma*	24	Myrtaceae	*Syzygium* sp. nov.	58
Verbenaceae	*Teijsmanniodendron bogoriense*	20	Burseraceae	*Haplolobus floribundus*	42
Fagaceae	*Lithocarpus sp.*	18	Myrtaceae	*Xanthomyrtus fusciculata*	41
Clusiaceae	*Calophyllum soulattri*	16	Lauraceae	*Cryptocarya idenburgensis*	25
Chrysobalanaceae	*Parastemon versteeghii*	15	Dipterocarpaceae	*Anisoptera polyandra*	24
Myristicaceae	*Gymnacranthera paniculata*	13	Myrtaceae	*Syzygium accuminatissima*	20
Burseraceae	*Haplolobus pubescens*	13	Verbenaceae	*Teijsmanniodendron ahernianum*	20
Myrtaceae	*Rhodomyrtus sp.*	12	Meliaceae	*Dysoxylum alliaceum*	16
Flacourtiaceae	*Ryparosa javanica*	12	Clusiaceae	*Garcinia hunsteinii*	16
Euphorbiaceae	*Neoscortechinia forbesii*	12	Chrysobalanaceae	*Parastemon versteeghii*	15

Table 4. Number of species, number of stems, and Family Importance Values (FIV, Mori *et al.* 1983) for the ten dominant families in the Alluvial and Hill plots.

FAMILY	ALLUVIAL PLOT # SPP. (%)	# INDIVIDUALS (%)	FIV	FAMILY	HILL PLOT # SPP. (%)	# INDIVIDUALS (%)	FIV
Euphorbiaceae	16 (8.8)	55 (9.6)	30.65	Myrtaceae	20 (12.8)	200 (26.4)	61.76
Lauraceae	15 (8.2)	37 (6.4)	22.55	Lauraceae	20 (12.8)	121 (15.9)	44.17
Myrtaceae	14 (7.7)	44 (7.7)	22.00	Burseraceae	3 (1.9)	51 (6.7)	19.43
Burseraceae	7 (3.8)	36 (6.3)	13.80	Meliaceae	11 (7.1)	41 (5.4)	18.46
Rutaceae	6 (3.3)	24 (4.2)	12.08	Clusiaceae	7 (4.5)	48 (6.3)	15.32
Fagaceae	2 (1.1)	27 (4.7)	11.77	Euphorbiaceae	8 (5.1)	28 (3.7)	13.16
Myristicaceae	8 (4.4)	25 (4.3)	11.48	Myristicaceae	8 (5.1)	24 (3.2)	9.94
Clusiaceae	5 (2.7)	30 (5.2)	10.03	Verbenaceae	2 (1.3)	23 (3.0)	9.46
Elaeocarpaceae	5 (2.7)	15 (2.6)	9.14	Sapotaceae	6 (3.8)	16 (2.1)	7.37
Sapotaceae	3 (1.6)	27 (4.7)	8.86	Chrysobalanaceae	2 (1.3)	16 (2.1)	7.04

Table 5. Dominant families from the hill and alluvial plots and other lowland sites in PNG (with Family Importance Values; Mori *et al.* 1983.)

SITE	IVIMKA ALLUVIAL	IVIMKA HILL	NAGORE SOUTH*	NAGORE NORTH*	SI RIVER*
Dominant families (FIV)	Euphorbiaceae (30.65)	Myrtaceae (61.75)	Meliaceae (36.11)	Moraceae (47.49)	Datiscaceae (42.09)
	Lauraceae (22.55)	Lauraceae (44.17)	Annonaceae (30.06)	Meliaceae (27.71)	Meliaceae (36.42)
	Myrtaceae (22.00)	Burseraceae (19.43)	Lauraceae (26.81)	Lauraceae (26.76)	Moraceae (31.17)
	Burseraceae (13.80)	Meliaceae (18.46)	Moraceae (26.22)	Sapindaceae (22.82)	Euphorbiaceae (18.98)
	Rutaceae (12.08)	Clusiaceae (15.32)	Sapindaceae (20.09)	Annonaceae (17.64)	Anacardiaceae (16.16)
	Fagaceae (11.77)	Euphorbiaceae (13.16)	Rubiaceae (14.44)	Euphorbiaceae (15.99)	Annonaceae (15.42)
	Myristicaceae (11.48)	Myristicaceae (9.94)	Nyctaginaceae (12.01)	Sapotaceae (11.34)	Myrtaceae (13.38)
	Clusiaceae (10.03)	Verbenaceae (9.46)	Euphorbiaceae (11.56)	Clusiaceae (10.43)	Combretaceae (10.98)
	Elaeocarpaceae (9.14)	Sapotaceae (7.37)	Sapotaceae (11.23)	Combretaceae (9.98)	Lauraceae (10.41)
	Sapotaceae (8.86)	Chrysobalanaceae (7.04)	Combretaceae (10.14)	Rubiaceae (9.83)	Sapotaceae (9.93)

*Data from Oatham and Beehler (1997).

Physical Parameters

The two plots differed in several characteristics beside the obvious difference in elevation (Table 6). The alluvial plot was essentially level whereas the hill plot traversed steep and varied topography. The subplots in the alluvial forest were more varied in canopy openness (Table 6), perhaps due to more or larger treefalls. The relative humidity of the hill forest was noticeably lower than in the alluvial forest; breezes unnoticeable in the alluvial forest cooled and dried the hill forest despite the relatively minor difference in elevation (personal observation).

Soil samples from the two plots exhibited some differences (Table 7). The color of soil from the hill plot was a homogeneous ferrous red, whereas in the alluvial plot, colors ranged from ferrous yellow to bluish gray. The surface of the soil in the alluvial plot was bare, and no horizon line was apparent between 0 - 50 cm depth. However, the hill plot had a surface layer 10-30 cm thick of mosses and leaf litter. Magnesium and sodium contents were both significantly higher in the alluvial soil samples, while copper content was significantly higher in the hill plot soil samples (Table 7). Soil characteristics and nutrient content of both plots were consistent with other tropical lowland silty clay soils (Gentry 1988). Generally nutrients, notably nitrate, ammonium, and phosphorus contents were low, while iron, and aluminum were high (Table 7). Cation exchange capacity (CEC) and pH were predictably low in both plots, and did not differ between plots.

Table 6. Characteristics of study plots.

PLOT	ELEVATION (M ASL)	LENGTH ORIENTATION	SLOPE	ASPECT	% CANOPY (MEAN)
Alluvial	100 - 120	069°	0° - 5°	none	55 -99 (82)
Hill	175 - 260	031°	25° - 50°	variable	70 - 95 (83)

Table 7. Soil analysis results from the Lakekamu Basin for nutrients (ppm), pH, and cation exchange capacity (CEC). The arithmetic means and standard errors of the means (SEM) are reported for the Hill and Alluvial plots. The test statistics from the Mann-Whitney U-test (T-statistic) and probability values (P) are reported for comparisons between sites. Bold faced P values indicate significant differences.

	HILL PLOT (SEM)	ALLUVIAL PLOT (SEM)	(T-STATISTIC)	P
NO_3	5.67 (0.52)	6.13 (2.16)	82	.161
NH_4	0.04 (0.03)	0.0 (0.0)	49	.463
P	0.0 (0.0)	0.71 (0.485)	60	.442
K	17.84 (9.93)	14.17 (2.13)	85	.083
Ca	7.51 (3.98)	9.37 (1.68)	51	.083
Mg	5.64 (1.95)	9.15 (1.59)	48	**.038**
Fe	167.77 (11.57)	126.92 (25.92)	58	.328
Mn	6.25 (3.17)	2.14 (0.69)	51.5	.083
B	0.17 (0.05)	0.11 (0.01)	74.5	.505
Cu	1.12 (0.21)	0.27 (0.05)	40	.002
Zn	0.73 (0.29)	0.44 (0.03)	70	.879
Mo	0.13 (0.13)	0.0 (0.0)	64	.721
Na	11.41 (0.35)	13.42 (0.69)	47	**.028**
Al	1198.79 (84.60)	1157.61 (62.47)	70	.879
pH	4.0 (0.05)	4.03 (0.05)	65.5	.798
CEC	6.75 (0.44)	5.80 (0.34)	81	.195

* All values for ammonium in the alluvial plot were zero, with the exception of one sample (25.13 ppm), which is not included in the analysis.

Discussion

Overall IRS Flora

Certainly two hectares are inadequate to describe the flora of the IRS study site. The steep increase of the species area curves (Appendix 1A) indicates that the woody flora was not completely sampled. The slight leveling of the species area curve for the hill plot (Appendix 1A) midway along the transect coincides with where the transect crested a ridge. This emphasizes the importance of randomly placing transects; a transect that followed a ridgetop might have underestimated the species richness of this site.

The plot data, combined with the general survey data (Takeuchi and Kulang, next section) yields a more complete picture of the flora. Neither method alone (plots or general collecting) produced a complete flora. The plot enumerations added eleven families and roughly 174 species to those found by general collecting (Appendices 2 and 3). Combined, the RAP survey revealed the presence of over 600 plant species in 130 plant families. Such high diversity apparent in such a small portion of the Basin in such a short time period substantiates the overall finding of the RAP survey that this is an area of rich biodiversity. It is indicative of the high diversity of the Basin to compare our results from two hectares: 253 species among 1,334 stems > 10 cm DBH, to the results of a fifty hectare plot in Panama: 303 species among 235,895 stems > 1 cm DBH (Hubbell and Foster 1992).

The creation of two marked plots with representative specimens in the nearby collections at FRI enhances the attractiveness of the IRS to scientists considering research in PNG.

Density, Dominance and Diversity

The hill forest had substantially more stems than the alluvial plot, but not disproportionately more in any size category. The hill forest plot had only a nominally greater total basal area and only two more very large trees (DBH ≥ 60 cm) (Table 2). Likewise, the stem size class distributions of both plots were remarkably similar (Figure 1).

Despite having more trees, the hill forest had lower diversity than the alluvial forest both in terms of alpha diversity and the Shannon Index (Table 1). The hill forest was dominated by two families—Myrtaceae and Lauraceae. These two families contained over 25% of the species and over 40% of the stems found on the hill plot whereas the alluvial plot had no such dominant families (Table 4). Interestingly, of the four families Takeuchi and Kulang (next section) list as dominants, three were dominant on the hill plot, but one they reported, Elaeocarpaceae, did not rate in the top ten hill families (Table 4) nor did the botanical survey team report Lauraceae as a dominant family. This emphasizes the importance of combining quantitative and qualitative assessments.

Both plots exhibited the pattern expected in tropical rainforests: many rare species and a few fairly common species. More than forty percent of the species on both plots were only represented by a single individual. Over 86% of the species on the alluvial plot were represented by 5 or fewer individuals compared to 64% on the hill plot. The higher diversity of the alluvial forest was characterized by a greater number of rare species. Whereas most species were represented by rare taxa, a substantial number of the stems were represented by a few common species. The top ten species comprised 26.9% of stems in the alluvial plot and 36% of the stems in the hill plot (Table 2). For the two plots combined, 7.5 % of the species accounted for 32.4 % of the stems.

Although the two Ivimka plots shared seven of the top ten families (FIV), they shared only 28.5% of their species (Appendix 2), shared only one domaint species (Table 3), and were strikingly dissimilar in terms of general floristics (next section). The two plots were separated by less than 5 km, had similar soils (Table 7) and only a small elevational difference (Table 6). Quite likely, the differences in ground slope (and hence drainage), and disturbance regimes (Table 6) between the two plots has a profound effect on vegetation. Conservation in the Basin that includes adjoining hill forest could preserve a vastly greater range of diversity in a relatively narrow fringe of surrounding hill forest.

CONSERVATION INTERNATIONAL

Physical Parameters

The soil samples were typical of poor tropical soils (Gentry 1988), with a Ca^+ content less than 100ppm. Both plots had very low pH, which leads to decreased bio-availability of phosphorus, calcium, copper and magnesium, especially in the presence of ferrous and aluminum oxides and calcium ions (Barber 1995). Phosphorus and nitrogen were both limited in our soils, suggesting that nutrients are very tightly cycled between detritus and live material. Low phosphorus content in the alluvial plot could limit litter fall, and contribute to the near absence of leaf litter in the alluvial plot (Vitousek 1984). The flat terrain in the alluvial forest apparently does not limit leaching rates relative to the steeper hill forest (where the moss layer might help slow erosion). Only magnesium and sodium were significantly higher in the alluvial plot and copper was significantly lower (Table 7). CEC was low, which limits plants' abilities to uptake nutrients. Quite likely, the native flora is well-adapted for low nutrient soil but exotic taxa (e.g. oil palm plantations proposed as a development project) would not flourish in the Basin. The Basin's poor soils could also limit or impede forest regeneration after logging.

During the one-month period of data collection, we observed eight treefalls in and around the alluvial plot. Treefalls are common and often large in the study area (B. Gamui and A. Sitapa, unpublished data). This creates heterogeneity of seral stages in the alluvial forest and probably contributes the elevated high diversity of the alluvial forest. Indeed our estimates of canopy cover varied more in the alluvial forest (Table 6), largely due to the presence of more canopy openings due to treefalls. Many factors, such as shallow, water-logged soils or high wind velocities could contribute to high treefall rates which in turn could limit tree size. Interestingly, the other plots in the Basin (Oatham and Beehler 1997) had more large trees (Table 1). It would be rewarding to compare disturbance rates at different sites in the Basin and the effect on tree size and diversity.

Vegetation in the Lakekamu Basin

Oatham and Beehler (1997) studied three one-hectare plots in the Nagore-Si region of the Lakekamu Basin, and reported a range of 392 - 482 woody stems per plot (Table 1). Due to the inclusion of lianas in their study, these data are not directly comparable to ours. However, as both Ivimka plots had more stems and more species (except the hill plot which had fewer than the Nagore South plot), the exclusion of lianas from the Nagore-Si plots would only widen the differences (Table 1). The greater basal areas and mean DBH at the Nagore-Si plots is probably not due to the inclusion of lianas, but rather the greater number of large trees. The Nagore-Si plots had more than twice the number of trees ≥ 60 cm DBH than on the Ivimka plots (Table 1).

Although the complete species composition data of the other three Lakekamu vegetation plots were not available, it is obvious the species composition of the three Nagore-Si plots differed substantially from the two Ivimka plots. Only three families were in the top ten in terms of family importance values in all five plots (Table 5), and eight were in the top ten in just one of the five plots. Among the top ten in each plot, there is no consistency in rank. Datiscaceae is the dominant family at the Si River plot and not in the top ten on any of the other four plots, Meliaceae ranked high in the Nagore-Si plots but was not in the top ten on the Ivimka alluvial plot.

Oatham and Beehler (1997) emphasized the heterogeneity among the Nagore-Si plots. Addition of the two Ivimka plots further emphasizes the heterogeneity in the Basin. None of the 23 top species in the Nagore-Si plots (Oatham and Beehler 1997) rank in the top ten in this study (Table 3). Furthermore of the 23 top species in the Nagore-Si plots, only three occurred (but were not dominant) in the alluvial plot and only one occurred in the hill plot. If we assume that undetermined species of one plot were equivalent to undetermined congeners in the other plot there could be as many as 12 of the dominant 23 Nagore-Si species on the alluvial plot and eight on the hill plot.

The Nagore-Si plots and Ivimka hill and alluvial plots all exhibited dramatic differences in species composition and structure. The two most separated sites are only about 20 km apart in continuous, closed forest. From the air, the forest of the Basin appears to be a complicated mosaic of different forest types (A. Mack, personal communication). Given the spatial variety of forest types in the Basin, it is clear that a successful conservation initiative will require coverage of most or the entire Basin. The causes and means of maintaining this heterogeneity are still speculative. Revealing these relationships would greatly assist conservation planning throughout New Guinea. Future research that combines remote sensing, by satellite or aircraft with additional ground survey of plots would be richly rewarding. The five one hectare plots in the Basin create a good beginning for such an initiative.

Comparison to Other Rainforests and Conservation Implications

Because the hill forest had some taxa characteristic of lower montane forests (*e.g.*, Podocarpaceae, Winteraceae), it is interesting to compare to site at the transition from hill to lower montane forest 250 km away in the Crater Mountain Wildlife Management Area (CMWMA) at 900 m elevation (Wright *et al*. 1997). The CMWMA plot had fewer stems per hectare, but substantially more species (Table 2). The Lakekamu plots are not as diverse as the CMWMA plot and all but the hill plot had substantially fewer stems. The two Ivimka plots each shared five top families with the CMWMA plot (Wright *et al*. 1997).

From a partial survey of rainforest vegetation studies (Table 2) the Ivimka plots lie within the range of values for species richness, number of stems and basal area per hectare. The basal areas in the hill (32.04 m^2) and alluvial (28.46 m^2) plots are somewhat below the estimated pantropical mean basal area (36 m^2, Dawkins 1959), but the tree density on the hill plot is somewhat high (Table 2). A large number of small trees might make these forests less attractive to commercial loggers, though the threat from logging is real. Of nineteen major timber export species (Sekhran and Miller 1995), at least seven occurred in the hill forest. However, only four occurred in the alluvial plot and only three timber species had ten or more stems ≥ 10 cm DBH on the pooled Nagore-Si plots (Oatham and Beehler 1997). Thus it appears the hill forest adjoining the Basin might be attractive to commercial loggers, but the timber in the Basin itself is probably less valuable. The shallow rivers of the Basin and road-building in the muddy, often-flooded alluvial Basin could make timber extraction difficult in the Basin.

Nearly one quarter of the world's lowland tropical rainforests is already gone with much of the rest of it existing in only small fragments (often on the order of 1 km^2; Turner and Corlett 1996). Conserving the remaining large, undisturbed areas of the world must be ranked among the highest priorities in biological conservation. The Lakekamu Basin is one outstanding candidate for conservation. The data indicate extraordinary diversity and heterogeneity within the Basin. The Basin comprises a logical conservation unit—a major watershed that is currently sparsely inhabited and devoid of roads or other major development projects. The area may have some valuable timber. However the data suggest timber species are scattered in their distribution and often small in stature, making extraction expensive especially in light of the limited options for road-building or log-rafting in much of the Basin.

The pristine quality of the Lakekamu ecosystem, its high diversity, its ready access by small aircraft and the newly-constructed Ivimka Research Station present an ideal environment for expanding research and possibly ecotourism in the area. The permanent plots, voucher plant series and other assets contributed by the RAP team should increase the area's attractiveness to scientists and tourists.

CONSERVATION INTERNATIONAL

Rapid Assessment Program

Figure 1. Size-class distributions of trees in the alluvial and hill plots. Size classes begin with trees ≥ 10 cm DBH in each plot. Labels indicate the upper limit of each size class. The size class distributions for the two plots did not differ statistically significantly.

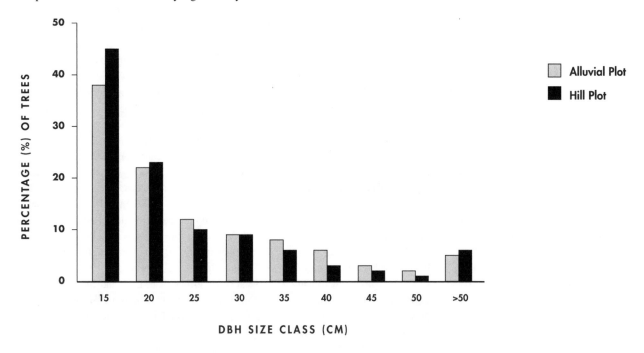

Figure 2. The percent of total species within a plot according to abundance classes. Abundance classes numbers of individuals ≥ 10 cm DBH representing a species within each plot. The two distributions did not differ significantly.

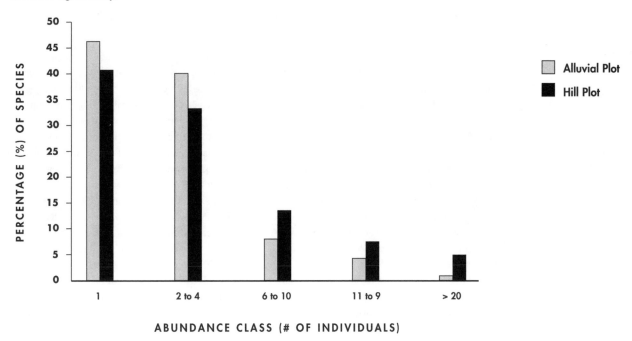

VEGETATION PART 2: BOTANICAL SURVEY (Wayne Takeuchi and Joel Kulang)

Chapter Summary

- About 450 species and morpho-species of vascular plants were recorded during the survey.
- Alluvial forest in the Basin is very species rich and comprised of a heterogeneous mosaic of different communities and various seral stages, with no strongly dominant species.
- The hill forest adjoining the Basin is less diverse than the alluvial forest and contains a suite of about ten dominant species.
- The frequency of natural disturbances in the alluvial forest, from windthrows and changes in river course contribute to the high diversity, heterogeneity and lack of dominance in the Basin, whereas the greater stability in the hill forest leads toward the opposite.
- There is an anomalous component of normally montane species descending to low elevations where they interface with the alluvial forest.
- The flora of the area has a strong component of local endemism and taxa confined to the Papuan (southern PNG) subregion; these taxa are poorly known, making the RAP collections particularly significant.
- There is little evidence of anthropogenic disturbance and few weedy, large-disturbance specialist species. The Lakekamu Basin has pristine wilderness vegetation, a condition that will soon be very rare globally.

Introduction

A botanical reconnaissance of the IRS study area was conducted between Oct. 14 and Nov. 12, 1996 by the principal author and two survey trainees. The team's primary responsibilities included species checklisting, collection of herbarium specimens, and *ad libitum* exploration. Our taxonomically-oriented activities were designed to complement the vegetation plots censused by a counterpart team and generate a preliminary floral description of the area.

Methods

Exploration and collections were made in all directions from the Ivimka Research Station sited on the Sapoi River. The reconnaissance followed established trails and the many stream channels in the study area. We spent 19 days working in alluvial and riverine forest and 9 days in the hill forest. Gatherings of fertile specimens were obtained for taxonomic study and field-pressed in 70% surgical alcohol for subsequent processing and determination at the Lae National Herbarium. Materials for exsiccatae were accompanied by numerous bottled, carpological, and xylarium accessory collections. Specimens were typically secured in multiple sets (up to 10) for principal distribution to Lae Herbarium (LAE), Botanical Research Institute of Texas (BRIT), Kew, Harvard University Herbaria, and Leiden. Residual duplicates will be distributed in accordance with the exchange protocols of LAE and BRIT. Common taxa already well-known to the lead botanist were simply checklisted as sight records.

Results and Discussion

Vegetation Description
Appendix 3 is a compilation of all sight enumerations and of species documented by actual collections. A total of about 450 morphospecies was recorded among vascular plants within the investigated elevational range from 100 m to 370 m.

From daily reconnaissance and collecting, a general impression was obtained of the vegetation around the base camp that corroborates the general findings from the vegetation plots. At the most basic level, there is a very obvious distinction between the alluvial forest on the Avi Avi-Sapoi flood plain and the foothill forest on the slopes above. The differences between the formations are structural as well as compositional. In comparative terms, the hill communities have lesser understory development and canopy stratification. Although there is noticeable intermixing of plant taxa, the hill forest seemed decidedly less rich than alluvial communities. At least on a local basis, hill canopies often exhibit frequency dominance by a suite of about 10 species (primarily a

Calophyllum complex with *C. euryphyllum, C. goniocarpum,* and *C. papuanum, Garcinia* sp., *Teijsmanniodendron ahernianum, Syzygium* spp., *Elaeocarpus sepikanus* and *E. blepharoceras*) and evoke a perception of homogeneity relative to the alluvial forest, where dominance relationships were not particularly apparent.

On the recent Hammermaster and Saunders (1995) forest classification, the Lakekamu mature-growth alluvial forests are primarily assignable to large to medium crowned forest, and open forest (structural codes Pl and Po respectively). Both forest types are low altitude (<1000 m) formations on plains and fans. Seral communities within the survey area are consistent with mixed riverine successions (code Fri). This latter category is represented by heterogeneous forest with many seral stages, and species composition is typically highly variable. Changes in streamcourse along meandering rivers are the major factor determining the character of such communities. Lowland forest classes Pl and Po are currently under consideration by the PNG Dept. of Environment and Conservation (DEC) for designation as logging exclusion zones, on account of the habitat value and damage susceptibility to logging of such forests (DEC 1996). The forest type represented by the Lakekamu successional communities are not regarded as requiring any sort of special designation or protection (ibid).

A conspicuous aspect of the Lakekamu regrowth phase is the collective rarity of plants usually common in disturbed environments. Taxa such as *Gouania, Mussaenda, Homalanthus, Commersonia, Melochia, etc.,* ordinarily prolific in anthropogenic habitats were typically present only as isolated individuals in the surveyed area, even in otherwise disturbed sites such as windthrows and riverbanks. Their infrequent occurrence suggests that disturbance conditions are spatially restricted and ephemeral, so that the recovery process ensues rapidly and the seral phases are quickly reset. Under such conditions, the weedy species characteristic of repetitively disturbed habitats never have an opportunity of becoming spatially dominant. When considered in conjunction with the structural characteristics of the forest, the image which emerges of this vege-

tation is that of a highly dynamic flora moving rapidly through maturational stages. The numerous treefalls and canopy opening events observed during the expedition support the notion of a latent ecosystem conditioned by unpredictable events.

The Lakekamu hill forest can be considered primarily medium-crowned forest on low altitude uplands (<1000 m) under Hammermaster and Saunders forest code Hm. Such communities are characterized by gradual transitions in floristic composition, with the interval below 500 m resembling large-to-medium crowned and open-forests (types like Pl and Po), and the elevations above 500 m being more similar to montane formations. This is probably the situation at the survey sites. However a significant permutation at Ivimka is the otherwise unexpected occurrence of montane genera such as *Rhododendron, Dimorphanthera, Rapanea, Zygogynum, Levieria, etc.* at descending elevations to the interface with alluvial plain. These seemingly anomalous occurrences may be indicative of some peculiarity in the Quaternary vegetational history of the Lakekamu flora, or alternatively, it may indicate just how little we know about the real ecological amplitude of these taxa.

In both the alluvial and hill environments, tree stocking densities and canopy heights varied considerably irrespective of forest type. Heavy rains were experienced during most of the expedition, but the amount of epiphytic growth was much less than what would otherwise be expected from such an apparently high-rainfall habitat. The fern and orchid flora was actually somewhat depauperate of species. Although hill substrates were consistently overlain with litter accumulations much thicker than the alluvial counterpart, this was likely a seasonal distinction rather than a characteristic structural feature.

The Lakekamu alluvial zone is clearly a composite of floristically differentiable communities. Forest sections immediately adjacent to rivers have species frequencies distinct from interior stands, in spite of the commonly shared taxa. Even within the forest proper however, there is discernable fragmentation in community structure. The large number of windthrows seen during

A conspicuous aspect of the Lakekamu regrowth phase is the collective rarity of plants usually common in disturbed environments.

the survey indicate that seral clusters of even-aged cohorts are typically establishing in scatter-shot fashion throughout the forest matrix. With maturation of the gap communities, a rotating spatial mosaic is collectively composed by the various cohort populations, each in its own particular dynamic phase.

The variation in spatial scale of community organization (as evidenced from tree densities, canopy development, species composition, *etc.*) suggests the parallel existence of disturbance mechanisms of complementary scale to the floristic variation. At one end of the spectrum are small-area successions resulting from individual windthrows and the breakage of large tree limbs. However, storms of progressive intensity would be expected to produce gaps of increasing size and continuity. Large shifts in streamcourse could also produce area-extensive displacements. The forest section along the trail/trapline opposite the 3k transect might be an example of such larger-scale past disturbance. Over time, the sequential occurrence of forest upsets with differing extension would be expected to superimpose on each other in random fashion. The cumulative result would be a heterogeneous forest with a confusing successional mixture of component parts. This may be the etiology for the Lakekamu alluvial mosaic, especially if substrate patterns cannot be linked to the aboveground patterns.

In contrast to the flood plain vegetation, hill forest within the surveyed area (to 370 m elevation) does not show signs of dynamic turnover to the same extent as the flatland. This comparative stability is accompanied by an apparent reduction in overall richness.

Supposed distinctions in forest quality derived from casual observation need to be confirmed empirically with systematic sampling. The two one-hectare plots of the RAP survey corroborate many general impressions, but more extensive sampling is recommended. A study program preceded by extensive reconnaissance and community entitation is suggested, to be followed by central placement of various releves or hectare plots within the entited units. Community separations could be thereby defined more closely, though the dynamic relationship between forest units can only be resolved by long-term tracking.

Phenological Patterns

Mass and synchronized flowering was noted during the survey for taxa in *Barringtonia, Calophyllum, Dimorphanthera, Dysoxylum, Garcinia, Grenacheria, Metrosideros, Neisosperma, Pternandra, Syzygium,* and *Timonius.* Periods of peak flowering and anthesis were typically very brief, sometimes lasting only a few days. Most plant species exhibited erratic flowering phenology, with populations exhibiting a full range of unsynchronized states. However even with the apparent randomly-behaving taxa, a commonly seen pattern was the occurrence of flowering trees in clusters, though conspecific individuals short distances away might be entirely sterile.

Extensive germinant and seedling crops were noted for *Calophyllum* in hill forest. With the exception of *Calophyllum,* mature understories were largely clear of germinant flushes, suggesting that canopy opening is otherwise necessary for germination of tree taxa.

Phenological patterns were difficult to assess on an impressionistic basis due to the selectivity of human observation. Attention is naturally drawn to fertile specimens, so the perceived proportion of fertile individuals from a population can be easily overstated. Objective sampling protocols need to be established to provide reliable data for ecological applications such as animal-plant interactions.

Botanical and Conservation Significance of the Lakekamu Tract

Botanists on rapid-format surveys work at a unique disadvantage relative to their zoological-study counterparts. Faunistic enumerations can occur by visual or auditory record, and captures need not be gravid to be identifiable but botanical specimens typically require fertile structures in order to be reliably named. This distinction applies especially to species-rich plant groups. Unfortunately, the large families are exactly where botanical records of greatest conservation and biological significance are likely to be secured.

The percentage coverage which the survey has achieved of the Lakekamu flora is of course unknown. Many taxa with unfamiliar vegetative features were seen throughout the survey, but could not be collected in fertile condition. Our gatherings probably encompass only a minor percentage of the aggregate flora, but it is worth noting that a substantial number of new taxa and records were nonetheless obtained. If such a comparatively small sample can yield significant results, the potential for further discovery is certainly even greater. The best finds have yet to be made, and there is much scope for future exploration.

Approximately 400 exsiccatae numbers were made during the RAP survey, and the most significant discoveries have been summarized in Appendix 4. Plants previously unknown to science, rare taxa, distributional and range extension records have all been documented by the collections.

A curious aspect of the discoveries from Lakekamu is that most of the new records represent taxa common in the survey area. This circumstance is a reflection of the low botanical collection density for PNG in general and for the Papuan territories specifically, a situation allowing even high-visibility taxa to remain undetected. At Lae Herbarium, the number of specimens from Gulf Province is among the lowest from any district. In the post-Independence period, concentration of the national infrastructure in forestry and botany to Morobe Province has indirectly contributed to a comparative neglect of the Papuan flora. The selection of Lakekamu as a survey site was thus especially appropriate, and further efforts directed to the southern side should continue to produce disproportionate returns.

Ecologically, the Lakekamu tract shows considerable promise as a research venue due to its natural-growth status and the rich mixture of closely juxtaposed seral communities. This area would be very suitable for investigations into the dynamics of the Papuan lowland forest and for evaluating connections between environmental latency and floristic diversity. The unusual elevational occurrences are also rather provocative and deserving of further inquiry.

The Lakekamu flora could support intensive taxonomic study on a number of plant groups. From impressions obtained during the survey, families Annonaceae, Euphorbiaceae and Clusiaceae (Guttiferae) are appropriate subjects for specialist examination, being composed of diverse and numerically prominent representatives. Generic limits are inadequately defined in Annonaceae (only Sapotaceae and Lauraceae are in comparable disarray), and the species rich Lakekamu assemblage would serve as an excellent subject for detailed study.

INSECTS PART 1: THE SOCIAL HYMENOPTERA (Roy R. Snelling)

Chapter Summary

• Social Hymenoptera (Apidae, Formicidae, and Vespidae) were surveyed 20 October to 8 December, 1996 using a variety of techniques (active searching for individuals and nests, Malaise traps, litter sifting, and baiting).

• A total of 281 species of social Hymenoptera were collected, represented by approximately 3000 specimens; 254 of these species belong one family, the Formicidae (ants).

• About twenty-two species collected represent new, undescribed species and one represents a new genus.

Introduction

The social Hymenoptera (ants, bees, and wasps) are an influential part of the biotic environment in New Guinea, as they are in most tropical rain forests. In both absolute numbers and biomass the social Hymenoptera, especially ants, often dominate arthropod faunas of tropical rainforests, both in the canopy and on the ground (Davidson 1997; Wilson 1987). It is estimated that about one-third of the entire animal biomass of the Amazonian *terra firme* rain forest is composed of ants and termites and that, along with bees and wasps, these insects comprise more than 75% of the total insect biomass (Fittkau and Klinge 1973). Other studies have suggested that much the same is true in African rain forest. While comparable studies have yet to be made in New Guinea, it is probable that similar results will be forthcoming.

At the bottom of the food chain, ants, bees, and wasps represent an almost unlimited, renewable, highly nutritious protein resource for a great many species of insectivorous vertebrates, from frogs and snakes to birds and bats.

Ants and social wasps are the chief predators on insects, spiders and other arthropods and, less commonly, on small vertebrates. The larvae of both groups are largely, if not entirely, carnivorous. Since colony populations of some species of ants are often in excess of 50,000 larvae, it follows that considerable quantities of insect prey are collected by the foraging workers in order to feed these larvae.

Ants ". . . form the cemetery squads of creatures their own size, collecting over 90 percent of the dead bodies as fodder to carry back to their nests. By transporting seeds for food and discarding some of them uneaten in and around the nests, they are responsible for the dispersal of large numbers of plant species. They move more soil than earthworms, and in the process circulate vast quantities of nutrients vital to the health of the land ecosystems" (Hölldobler and Wilson 1994).

The social Hymenoptera, together with the termites but in marked contrast to nearly all other terrestrial invertebrates in the Lakekamu area, constitute a year-round source for study. Most other invertebrates are more or less seasonal: part of the year only adults are present and part of the year only larvae are present. Such seasonality makes it difficult to accurately survey a fauna in a short period of time. For this reason, the social Hymenoptera, with their stationary, perennial nests are excellent survey subjects.

Some groups of insects, especially those that include conspicuous or "showy" species are reasonably well collected and well known (*e.g.*, butterflies and certain families of beetles). However, the less gaudy insects are less often and less intensively collected, hence much less well known. The New Guinea Hymenoptera, despite their ecological importance, are relatively poorly known. There are no comprehensive accounts of any of the families including social species for New Guinea as a whole, or for any significant portion of the island.

Methods

Non-ants

Apidae and Vespidae were collected principally by two methods: the netting of foraging individuals and the capture of entire colonies. Since Polistinae, the most species rich subfamily of Vespidae, most frequently nest on the undersides of leaves or on branches of low-growing shrubs, their nests were often discovered by other researchers and reported to me. I collected complete colonies whenever possible.

Six Malaise, or flight-intercept, traps were deployed at selected sites along the Bulldog Road and along the Ridge Trail (see map page 22). These traps collect flying insects directly into containers of alcohol and are an excellent method for sampling the fauna of a small area. The traps were placed in various situations: closed canopy forest, clearings, and at the edge of clearings in an effort to secure as great a diversity as possible. Traps were serviced at ten day intervals and their contents stored for later examination.

Ants

Samples were obtained by direct examination of rotting logs and branches, searching under loose bark, under stones, splitting stems and twigs of living plants, examination of such myrmecophytes as *Myrmecodia* (Rubiaceae) and *Endospermum* (Euphorbiaceae) and epiphyte root masses, and by searching for foraging columns and individuals.

In addition, litter samples were taken. The coarse leaf litter was sifted out; the resultant accumulation of fine litter and duff was then divided among several coarse-meshed bags and placed in a collecting bag. As the litter dried out, the arthropods contained therein dropped out of the mesh bags into a container of alcohol at the bottom of the collecting bag. This method enables the collection of minute (less than 1.5mm long), cryptic, or lethisimulating species inhabiting the leaf litter that are easily overlooked by direct examination.

Pitfalls were not used as a survey method. The results of limited pitfall traps set out by a group of students included only a few of the most common surface-foraging species. Canopy sampling was limited to recent (< 3 days old) treefalls and was, therefore, strictly opportunistic. Ants at older treefalls relocate to adjacent trees. Only one recent treefall was examined during my stay at Ivimka, but it yielded several species not otherwise seen.

A primary set of voucher specimens (including types) from the Ivimka camp survey will be deposited in the Entomology Section of the Natural History Museum of Los Angeles County. Additional vouchers will be deposited, to the extent that duplicates are available, in the following institutions: Australian National Insect Collections, Canberra; B. P. Bishop Museum, Honolulu; Museum of Comparative Zoology, Cambridge, MA; The Natural History Museum, London; University of PNG, Waigani.

Results

The complete species list of social Hymenoptera collected on the RAP survey is given in Appendix 5. Additional specific results are discussed below. Annotations and natural history notes on the social Hymenoptera of the Lakekamu Basin are given in Appendix 6. Although the rate of collecting additional species slowed toward the end of the survey, new species were still being collected up to the last day (Appendix 1B), indicating continued sampling would undoubtedly reveal more species in the area.

Apidae (Social Bees)
Only four species of social bees were encountered at Ivimka, the Asian honeybee (*Apis cerana*) and three species of "sweat" or "stingless" bees (*Trigona* spp.). All were common and routinely attracted to perspiring humans or any other source of salts.

The New Guinea bee fauna is poorly studied, even for such relatively common and conspicuous bees as the *Trigona* species. At present there is no informed estimate as to the number of *Trigona* species present in New Guinea. At least some of

the "Indo-Malayan" species treated by Schwarz (1939) are also present on New Guinea, as presumably is true of some now known from the Cape York Peninsula of Australia. Michener (1965) lists 10 species for New Guinea, but there are probably more.

Formicidae (Ants)
A total of 266 samples of ants was made, 254 of which represent individual collections; the remaining 12 samples are from sifted litter. Two hundred and fifty-four species of ants belonging to 59 genera were collected at Ivimka, a much higher number than I had anticipated. In an area of lowland rain forest at the lower Busu River, Wilson (1959c) found 171 species in 51 genera. Wilson's study is the only prior base-line New Guinea work available for comparative purposes; the Busu River fauna examined by Wilson is notable for its Melanesian components. Wilson commented that "the ant fauna of the Busu-Bupu area is perhaps the richest ever recorded for a single locality anywhere in the world . . . it has been estimated that at least 59 genera and 172 species occurred within a few square kilometers in the collection area." [More recent systematic work has reduced Wilson's 59 genera to 51]. The results of Wilson's study and mine are compared in Table 1.

In the list of the ants collected at Ivimka (Appendix 5, Table 1), some of the unidentified species are represented by alates (winged individuals) only. Except in the case of the several *Aenictus* species, these are winged females, usually attracted to lights at night. It is usually possible to associate females with their respective worker forms on the basis of similarities in sculpture and pilosity. In the absence of clear indications of likely associations, some of these females have been listed here as separate species. These belong to genera that include mostly high-arboreal species and are assumed to represent additional species.

Wilson (1959c) described the ecological stratification of the ants of the lower Busu River. He recognized three reasonably distinctive strata:

1) The *Ground stratum* comprises those species that nest in the soil, leaf litter, and all rotting wood on the ground, up to and including the largest rotting logs. The majority of both species and genera live at this level and relatively few species leave it to forage in the arboreal zones.

2) The *low arboreal stratum* includes species that nest in herbaceous and shrubby ground vegetation and up to the first several meters of the trunks and branches of larger trees. A small number of species nest primarily or exclusively in this zone. Species nesting in this stratum commonly forage down to the ground stratum and some species from the ground stratum forage into this level.

3) The *high arboreal stratum* species nest in the upper trunks and canopy of the A- and B-stratum trees; ants nest within epiphytes, abandoned termitaria, and preformed cavities in both living and dead wood, including twigs. Although most of these species confine their foraging activities to the high canopy, a few do forage all the way to the ground.

Ants may be further characterized by their general feeding habits. Thus some ants may be specialist predators (**S**), general predators (**G**), pastoralists (**P**), or seed harvesters (**H**) (see APPENDIX 5). As noted in Appendix 6, most species of the ponerine genus *Myopias* are presumed predators on millipedes; some species of *Leptogenys* are specialist predators, but no data are available for most of our species. Within the Myrmicinae, *Strumigenys* tend to specialize on Collembola, but apparently will capture other arthropods as well. Most of our species are general predators and scavengers, readily attacking living arthropods when feasible, often taking dead or dying prey, as well. These ants also will take other food items, including seeds, nectar, and other general scavenge. Pastoralists include species of *Acropyga, Philidris, Anonychomyrma*, and at least one species of *Podomyrma*. None of the species observed at Ivimka was confirmed to be a seed harvester, but it seems likely that one or more of the species of *Pheidole* may fall into this category.

Aenictinae

Wilson (1964) recorded seven New Guinea species of *Aenictus*, known only from workers; an additional two species, based on males only, were not treated. At Ivimka, workers of two species were collected. Based on the males, we have a minimum of five species at Ivimka. Because these male species cannot now be associated with workers, the present list is assumed to be inflated. Since at present we cannot match male ants with workers of many taxa, study of basic life histories of New Guinean ants is clearly needed.

Cerapachyinae

Wilson (1959b) recorded 11 species of *Cerapachys* from New Guinea, six of which were noted to be present at the Lower Busu River site (not four as stated by Wilson 1959c). At Ivimka, ten species were recorded. Of these, seven appear to be species new to science.

Dolichoderinae

At Ivimka, the subfamily Dolichoderinae includes only a few species of non-stinging ants, most of which are arboreal. Fifteen species were encountered at Ivimka, of which two are non-native. *Tapinoma melanocephalum* is a now tropicopolitan "tramp" species originally from Africa. *Technomyrmex albipes* is of Asian origin and now widespread from India to Australia and throughout much of Oceania; it is recently established in the United States (California). These are both "insinuators" that have apparently minimal impact on established native species.

Formicinae

This subfamily is among the three most species rich in the Ivimka area with 58 species in nine genera. The genus *Polyrachis* is the most species rich ant genus with 28 species recorded during the survey. The subfamily also includes the largest ant species found, *Camponotus dorycus*. Among the specimens collected on the survey are one undescribed species of *Echinopla*, one of *Paratrechina*, and three of *Polyrhachis*.

Myrmicinae

The Myrmicinae is by far the most species rich subfamily in the Lakekamu Basin with 103 species in 26 genera. Members of the subfamily represent a diverse array of habits, including predators to coccid-tending exudate feeders. At least seven collected species are new to science and at least one represents a genus new to science.

Ponerinae

The Ponerinae, together with the Aenictinae and Cerapachyinae, are among the better known of the New Guinean ant fauna, largely due to the several taxonomic studies by Taylor (1967) and Wilson (1958a,1958b, 1959b). Several difficult groups, especially the genera *Hypoponera* and *Myopias*, have yet to receive comprehensive taxonomic treatment. A total of 57 species in 17 genera were collected. At least two species collected are new to science. A single male specimen, collected in a Malaise trap, is evidently the first New Guinea record of the rare genus *Probolomyrmex*.

Pseudomyrmecinae

This subfamily is largely tropical, with two genera (*Pseudomyrmex* and *Myrcidris*) in the New World and one genus (*Tetraponera*) in the Old World. At Ivimka, four species were collected, none of which appear to be new to science.

Vespidae (Social Wasps)

All PNG social and subsocial wasps belong to the family Vespidae and are divided among three subfamilies, the Polistinae, Stenogastrinae and Vespinae (Snelling 1981). All three of these subfamilies are present at the Ivimka study site, with the cosmopolitan subfamily Polistinae clearly the dominant group in both species and general abundance. As has been noted by various authors (*e.g.*, Gressitt 1959, 1961), the insect fauna of New Guinea is largely Oriental in its affinities; the genera *Parapolybia* and *Vespa* are primarily Oriental and reach their southern terminus in New Guinea.

Three genera of Polistinae have been recorded from New Guinea: *Parapolybia*, *Polistes*, and *Ropalidia,* and all have been collected at Ivimka. *Parapolybia* has not been previously recorded from PNG, although our one species was common at Ivimka. Vecht (1966) described *P. varia* ssp. *furva* from a few specimens collected in the Kebar Valley, Vogelkop Mountains of Irian Jaya. Our specimens are similar in color pattern, both darker and less abundantly yellow marked than in the more "typical" form which ranges from India to China and south to Borneo, Sumatra, and the Philippines. This wasp constructs single comb nests on the underside of leaves; fully mature nests may contain up to about 200 workers. These wasps are moderately aggressive and capable of delivering a painful sting.

The most species rich social wasp genus at Ivimka is *Ropalidia* (represented by 15 species). Unfortunately, the systematics of the New Guinean fauna is not well worked out. The pioneering work of Vecht (1941, 1962) on the *Ropalidia* of the Indo-Australian region was never completed, leaving the large and complex New Guinean fauna essentially untouched. One paper by Cheesman (1952) is virtually worthless and Richards (1978) treated a few New Guinea forms in a cursory fashion. A thorough taxonomic study of the New Guinean representatives of this genus would be worthwhile. I estimate that there are perhaps as many as 50 species in New Guinea; with 15 species presently known from Ivimka, this fauna seems unusually rich and deserving of intensive study.

The subfamily Stenogastrinae is small with about half a dozen genera and fewer than 50 species. Various genera of stenogastrines range from India to China, south to New Guinea. The genera *Anischnogaster* and *Stenogaster* are represented in New Guinea by several species with one of each genus present at Lakekamu. C. K. Starr is currently revising the subfamily.

Table 1. Comparison of ant faunas at the Lower Busu River (Wilson 1959c) and Ivimka camp (this study).

GENUS	Number of Species Recorded	
	LOWER BUSU RIVER	IVIMKA STUDY AREA
AENICTINAE		
Aenictus	2	7
CERAPACHYINAE		
Cerapachys	6	10
DOLICHODERINAE		
Dolichoderus	0	1
Iridomyrmex[1]	5	8
Leptomyrmex	2	2
Tapinoma	0	2
Technomyrmex	1	1
Turneria	2	1
FORMICINAE		
Acropyga	3	4
Calomyrmex	1	1
Camponotus	8	13
Echinopla	2	3
Euprenolepis	0	1
Oecophylla	1	1
Paratrechina[2]	6	7
Plagiolepis	2	0
Polyrhachis	15	27
Pseudolasius	1	1
MYRMICINAE		
Adelomyrmex	1	1
Ancyridris	1	0
Aphaenogaster	2	0
Cardiocondyla	2	4
Crematogaster	6	11
Dacetinops	1	1
Dilobocondyla	1	1

GENUS	Number of Species Recorded	
	LOWER BUSU RIVER	IVIMKA STUDY AREA
Eurhopalothrix	0	4
Kyidris	1	0
Lordomyrma	3	3
Mayriella	0	1
Meranoplus	1	2
Metapone	0	2
Monomorium	1	3
Myrmecina	3	3
Oligomyrmex	2	4
Pheidole	15	13
Pheidologeton	1	1
Podomyrma	5	6
Pristomyrmex	3	6
Rhopalomastix	1	0
Rhopalothrix	4	1
Rhoptromyrmex	0	1
Rogeria	0	1
Smithistruma	0	2
Solenopsis	1	2
Strumigenys	12	16
Tetramorium[3]	2	9
Vollenhovia	2	4
Incertae sedis	3	1
PONERINAE		
Amblyopone	0	1
Anochetus	2	2
Cryptopone	2	2
Diacamma	1	1
Discothyrea	0	1
Gnamptogenys	1	1
Leptogenys	3	5
Myopias	6	8
Myopopone	0	1

GENUS	Number of Species Recorded	
	LOWER BUSU RIVER	IVIMKA STUDY AREA
Odontomachus	3	6
Pachycondyla[4]	7	8
Platythyrea	1	2
Ponera[5]	6	12
Prionopelta	1	1
Probolomyrmex	0	1
Proceratium	0	1
Rhytidoponera	3	4
PSEUDOMYRMECINAE		
Tetraponera	3	4
GENERA	51	59
SPECIES	171	254

[1] Includes species now placed in *Anonychomyrma* and *Philidris* (see Shattuck 1992).

[2] Includes *Nylanderia* of Wilson (1959c).

[3] Includes *Triglyphothrix* and *Xiphomyrmex* of Wilson (1959c, see Bolton 1977).

[4] Includes *Brachyponera, Ectomomyrmex, Mesoponera,* and *Trachymesopus* of Wilson (1959c, see Bolton 1995).

[5] Includes species now placed in *Hypoponera* (see Bolton 1995).

Discussion

While the social bee fauna was within expected limits of abundance and diversity, the ants and social wasps far exceeded expectations. With respect to the ants, I can paraphrase Wilson's (1959) comment on the ant fauna of the lower Busu River: The Ivimka ant fauna is possibly the richest ever recorded for a single locality anywhere in the world! This is all the more impressive when one considers that all the material collected at Ivimka was taken no more than 50 meters (and mostly within 10 meters) from any trail and that the aggregate area surveyed was almost certainly less than one km². Equally impressive is that the acquisition of additional species never ceased; I was acquiring additional ant species up to the last collecting day at Ivimka, a virtual guarantee that not all the species available at that site had been collected (Appendix 1). Collecting in the forest canopy would undoubtedly increase the list of species present in the area.

Similarly, the social wasp fauna seems to be extraordinarily rich, at least within the Old World tropics. This is no doubt due, at least in part, to the fact that no concerted surveys have been undertaken to determine the diversity of social wasps in any limited area of New Guinea. From 5 June-15 July, 1996, I collected 17 species of social wasps in Hong Kong, and possibly 1-3 additional species might ultimately be found there. At Ivimka I found 23 species in a much smaller area.

Ivimka supports a rich and varied fauna of ants and social wasps, certainly the richest documented for any locality in New Guinea. In large part this diversity is due to the relatively undisturbed nature of Ivimka and the nearly complete lack of adventive species. Species exotic to the area are apparently unable to thrive in such habitats. In contrast, introduced ants predominate in Tekadu Village, a bare 12 km away.

Important follow-up studies at Ivimka should focus on, and monitor, the influence of non-native (adventive) ant species. I predict that so long as the area remains no more disturbed than it is at present, the few introduced species should have little or no impact on the native ant fauna, nor on other faunal or floristic constituents at Ivimka.

The current mix of non-native species consists entirely of "insinuator" species that occupy marginal niches and do not normally offer serious threats to native species so long as the area is minimally impacted by human activities.

However, great care must be exercised to assure that potentially more destructive species, such as Fire Ants (*Solenopsis geminata, S. wagneri*), do not become introduced into the Basin. These polygynous, highly aggressive New World species not only compete successfully with native ants, they often may be a threat to nestling birds, especially those on or near the ground; similarly, immature mammals may also be endangered. *S. geminata*, in particular, can be a problem for two reasons. First, it is already widespread in the Oriental and Melanesian regions. Secondly, it is a tropical forest species, hence pre-adapted to the conditions existing at Ivimka and, indeed, throughout lowland areas of New Guinea.

Although not as conspicuous as vertebrates, ants are a dominant element in any tropical ecosystem and the overall health of any given habitat is dependant on the ants resident there. However, the role of the many species and/or feeding guilds is poorly understood. Although I managed to collect at least a dozen species of *Pheidole*, nests of most were not located and I do not know which, if any, are seed-gathering species. Almost two dozen species of *Polyrhachis* are present at Ivimka. Of their biologies, almost nothing is known. The ants at Ivimka could provide a study resource for ecologists for years to come.

The Ivimka ant fauna is possibly the richest ever recorded for a single locality anywhere in the world!

INSECTS PART 2: ODONATA (DRAGONFLIES AND DAMSELFLIES)
(Stephen Richards, Miller Kawanamo and Geordie Torr)

Chapter Summary

• We collected 34 species of odonates: 17 anisopterans (dragonflies) and 17 zygopterans (damselflies).
• One of the damselflies is a distinctive unde-

scribed species of uncertain generic affinities and one of the dragonflies appears to be identical to *Diplacina erigone*, a species previously known only from Misool Island.

• Several other species in the genera *Idiocnemis, Nososticta*, and *Drepanosticta* are almost certainly undescribed.

• We present a checklist and summarize the habitat preferences of each species.

Introduction

The forested lowlands and foothills of New Guinea south of the central mountain range remain poorly surveyed for odonates. Important collections were made on the southern slopes and lowlands of the Snow Mountains in Irian Jaya by the Lorentz expedition of 1909 (Ris 1913) and the Wollaston Expedition and British Ornithologists Union Expedition during 1912-13 (Campion 1915). Unfortunately, the first Archbold expedition which surveyed between Yule Island and the summit of Mt. Albert Edward east of Lakekamu in 1933-1934 concentrated primarily on obtaining specimens of vertebrates and plants (Archbold and Rand, 1935).

Within the past two years two major odonate collections have been assembled in the southern watersheds of New Guinea; during an extensive survey of the Kikori River Basin in Southern Highlands Province and Gulf Province, PNG (Polhemus 1995), and along an altitudinal transect across the southern slopes of the Snow Mountains in Irian Jaya (Polhemus 1997). The odonate fauna of lowland rainforests in southern PNG east and west of the Kikori Basin remains virtually unknown. This survey in the Lakekamu Basin provides the first data on odonates from this large unstudied area.

Methods

We conducted a survey of Odonata for 23 days (14 November to 6 December 1996) in lowland alluvial rainforest (120 m asl). The area surveyed was one kilometer wide and extended four kilometers south from the Ivimka Research Station along the Bulldog Road, adjacent to the Sapoi

River which formed the western boundary of the survey area.

Collecting Protocol

Specimens were collected with insect nets and by hand. Collecting was largely opportunistic; no attempt was made to standardize collecting effort in different habitat types, to accurately determine abundance, or to ensure that collecting effort was spread evenly over daylight hours. However, during the survey we collected extensively in each of the five recognized habitat types (see below), in a range of weather conditions, and at various times of the day including early morning and late evening (to sample crepuscular species).

Voucher specimens of all species except *Ictinogomphus lieftincki* were collected, stored in paper envelopes, and deposited at James Cook University, Townsville, Australia and in the UPNG insect collection.

Habitats

We recognized five distinct habitat types, based on broad patterns of water flow and availability, and sunlight penetration:

1) River. The only river surveyed was the Sapoi. This permanent watercourse is approximately 15-20 m wide and has clear, turbulent water and numerous emergent boulders embedded in a sand and rock substrate. The width of the river bed has resulted in reduced canopy cover and extensive exposure to sunlight for much of the day.

2) Streams. The study area is crossed by numerous small streams. These are permanent tributaries of the Sapoi River, are clear but slow-flowing, and generally <0.5 m deep. Most are 1-2 m wide and canopy cover is 100% so that in all but one site (where a treefall increased light penetration) only dappled light penetrates to the creek bed. The substrates are predominantly sand and small rocks.

3) Pond. A single pond, approximately 30 m X 8m was located in the study area, one kilometer south of Ivimka Camp. Canopy cover is 100% over the pond which is about 0.5 m deep when full, and has a deep substratum of rotting litter. The pond is temporary; it dried after about seven days of low rainfall during the survey, but filled again during the next heavy rain.

4) Clearings. There were several clearings in the survey area. These were of two types; a) those caused by natural treefalls and b) those formed by clearance of the forest for construction of the research facilities and helicopter landing pad. Standing water was generally not available in the clearings (one open-topped water tank was situated in the research hut clearing).

5) Forest. The forest understory itself, including the narrow trails formed during research activities. Construction of trails did not alter canopy cover, so light regimes were similar to those in undisturbed forest. Small depressions on the forest floor filled with water following exceptionally heavy rains but these usually dried within 48 hrs and were dry for most of the study period.

No attempt was made to sample canopy-dwelling species.

Results and Discussion

A total of 34 species were recorded during the survey (17 dragonflies and 17 damselflies); they are listed with their patterns of microhabitat use in Appendix 7. Because small streams meandered throughout the study area, many damselflies recorded as in "Forest" were relatively close to water. Only those perched immediately adjacent to, or over, water were classified as being "Stream" dwellers. Many of the species are poorly known, and were represented in the collection by less than five specimens. Although the habitat preferences reported here should be viewed as preliminary, it is clear that dragonflies (Anisoptera) occurred predominantly in forest clearings and along the Sapoi River (eight and ten species respectively) where light penetration is greatest, while most damselflies occurred along small shady streams (ten species) and in the forest understory (11 species) (Appendix 7).

Conspicuous exceptions were the large calopterygid *Neurobasis australis*, and the chlorocyphid *Rhinocypha tincta semitincta*, which were abundant along the sunny banks of the Sapoi River. The single specimen of an undescribed *Teinobasis* was also collected on a sunny bank of the Sapoi River.

In general, the dragonflies are representatives of common species that are widespread in New Guinea. Conversely many damselflies from Lakekamu represent significant range extensions of poorly known species, or are undescribed and possibly endemic taxa. As little information is readily available for odonates in PNG, we provide short species accounts in Appendix 8 in hope that future workers can improve our knowledge of odonates in the Lakekamu Basin.

The nymphs of odonata live in freshwater and are predatory, feeding upon small aquatic organisms and sometimes larger organisms like tadpoles. Many species have fairly narrow tolerances for water contaminants, turbidity and oxygen content. Additionally, the prey they depend upon may also have specific requirements. Thus, odonates can serve as useful indicators of water quality. With practice, adults of many species of odonates can be identified visually, allowing non-destructive and repeatable censusing. Dramatic changes in odonate populations can reveal changes in water chemistry that are otherwise difficult to detect by scientists monitoring water quality. Extractive industries such as agriculture, mining and commercial logging have serious potential impacts on freshwater quality; close monitoring of odonate and other aquatic organisms is an excellent method for assessing the effectiveness of management practices. The information collected here can act as a basis for continued monitoring of water quality as the LCI progresses. Additionally, data collected in the pristine environment of the Lakekamu Basin could prove useful in assessment of environmental impact of projects in other parts of PNG. We recommend future study of the freshwater systems in the Lakekamu include study of odonate natural history and population biology.

FISH (Gerald R. Allen)

Chapter Summary

• The Upper Lakekamu Basin fish fauna consists of 23 species belonging to 18 genera and 14 families.

• The fauna is dominated by relatively few families. Catfishes, rainbowfishes, gobies, and gudgeons are particularly prominent. This trend is typical of most freshwater localities in New Guinea.

• The number of fish species diminishes upstream with increasing elevation and current velocity.

• The Sapoi River in the vicinity of Ivimka Camp provides an excellent opportunity to study upstream species attenuation. The species total decreases from 19 to 6 over a distance of approximately 8 kilometers.

• The majority of Upper Lakekamu Basin fishes are distributed widely either across the southern portion of New Guinea or the combined northern Australia-southern New Guinea region.

• Two undescribed gobies belonging to the genera *Glossogobius* and *Lentipes* were collected during the survey, and a species of rainbowfish of the genus *Melanotaenia* is also a possible new species.

• The new *Lentipes* represents a significant find as the genus was previously unknown from New Guinea. The author recently discovered a second New Guinea species in the vicinity of Jayapura, Irian Jaya.

• There is probably no endemism in the fish fauna of the Upper Lakekamu Basin. The new *Lentipes* has not been collected elsewhere, but most likely occurs in adjacent drainage systems.

• Except for the apparently rare occurrence of the African Tilapia, the Upper Lakekamu Basin is uncontaminated by introduced species.

• Fishes form an important part of the Upper Lakekamu Basin biota and should form an integral part of any conservation programs that are implemented.

Introduction

This report contains a comprehensive documentation of the fish fauna of the Upper Lakekamu Basin, PNG based solely on a the RAP field survey conducted between 9 and 20 November 1996.

The principle aim of the survey was to provide a comprehensive inventory of the fish fauna of the Upper Lakekamu Basin, with special emphasis on species that might be unique to this area. The results of this survey should prove useful in evaluating the overall conservation significance of the Upper Lakekamu Basin. Survey results also permit direct comparison of the fish fauna with other river systems, both within and outside New Guinea.

Methods

A variety of methods was utilized in surveying the fish fauna, although rotenone was the primary means of procuring collections. This chemical is derived from the derris plant, and is ideal for collecting fishes in small creeks or sections of larger streams where current flow is minimal. The general method consisted of mixing approximately 0.5 - 1.5 kg of rotenone powder with several litters of water. This solution was then dispersed over a period of 5-10 minutes. After several minutes of exposure the stunned fishes begin to gasp at the surface and are easily netted. Rotenone stations were generally made in small creeks near their junction with larger streams. A representative, yet minimal sample of fishes are obtained in this manner, as the rotenone is quickly rendered inactive when diluted by the flow of the larger stream.

Two basic types of seine nets were also employed. The choice of net was dictated by habitat conditions. For narrow creeks, rivulets, and small, isolated rocky pools a one-person seine was used. This device consists of a one-meter square piece of fine-mesh netting attached between a pair of 1.5 meter-long poles. The net is weighted along the bottom edge and pushed along in front of the collector, as though pushing a lawn mower. This net also works well along grassy banks of larger streams. A 15 meter-long, fine-meshed seine was used in larger creeks and the edges of rivers. It has a width or "drop" of about 2.0 meters, and is weighted with lead along the bottom edge, and has a number of floats on the upper edge. The best technique for using this type of net is with one person at each end. One person holds a more or less stationary position on the bank while the other wades out into the stream and then returns along a roughly u-shaped path, then both ends are quickly hauled ashore.

A pair of small, hand-held scoop nets was effective for catching *Glossogobius* and the freshwater sole (*Synaptura*). The final method of collecting consisted of a rubber-propelled, multi-prong spear. It was used in conjunction with a diving mask and snorkel and proved indispensable for collecting fishes that live in areas of strong current such as *Lentipes, Kuhlia,* and *Cestraeus.*

Fish specimens were initially fixed in a 10 percent formalin solution and later transferred to 75 percent ethanol for permanent storage in museum collections. Most of the material was deposited at the Western Australian Museum, Perth, but representative collections will be lodged at the U.S. National Museum of Natural History, Washington, D.C., and the Zoological Museum, Amsterdam.

List of Collection/Observation Sites
(see also Appendix 9)
Station 1 - Ivimka Creek, about 200 km E of camp, 7°44.2'S, 146°29.9'E, elevation approximately 120 m, mud, sand and rock bottom, water clear with minimal flow through closed-canopy rainforest; mainly shallow with pools to 0.7 m deep, 0.5 kg rotenone (G.R. Allen, 11 November 1996).

Station 2 - Sapoi River at crossing of Bulldog Road, about 400 m upstream from Ivimka Camp, 7°43.9'S, 146°29.7'E, elevation approximately 130 m, bottom of cobbles, boulders, and sand, water clear with slow to very rapid flow through open forest, 3-m deep pool at base of shallow rapids, spear and seine (G. Allen and K. Merg, 12 November 1996).

Station 3 - Sapoi River, small side channel near "junction" with Avi Avi River, 7°43.5'S, 146°29.8'E, elevation approximately 140 m, bottom of cobbles, boulders, and sand, water clear with moderately fast flow through open forest, depth to 1.0 m, 1 kg rotenone (G. Allen, 13 November 1996).

Station 4 - Large (50 x 20 m) pool at junction of two creeks, about 3 km down Kakoro Track from Ivimka Camp, pool situated on edge of large garden clearing, approximately 7°45.3'S, 146°30.7'E, elevation approximately 120 m, bottom of cobbles and sand, water clear and slow-flowing through open forest, depth to 2.5 m, spear and one-person seine (G. Allen, 14 November 1996).

Station 5 - Sapoi River, small side channel about 4.5 km (straight-line distance) S of Ivimka Camp, 7°45.9'S, 146°29.0'E, elevation approximately 40 m, bottom mainly mud and sand with tree roots and logs, water clear and slow-flowing through open forest, depth to 1.0 m, 1 kg rotenone (G. Allen, 15 November 1996).

Station 6 - Small (1-2 m wide) forest tributary of Sapoi River, 70 m long section that empties into main river, about 1.6 km (straight-line distance) S of Ivimka Camp, 7°44.8'S, 146°29.3'E, elevation approximately 80 m, bottom mainly mud and sand with tree roots and logs, water clear and slow-flowing through closed-canopy forest, depth to 1.5 m, 0.5 kg rotenone (G. Allen, 16 November 1996).

Station 7 - Sapoi River, easternmost channel about 5.5 km (straight-line distance) S of Ivimka Camp, 7°46.6'S, 146°29.0'E, elevation approximately 35 m, bottom of cobbles, mud, and sand with tree roots and logs, water clear and slow-flowing through open forest, pool with depth to 3.5 m, 1.5 kg rotenone (G. Allen, 17 November 1996).

Station 8 - Sapoi River, "Turtle Pool" in easternmost channel about 4.6 km (straight-line distance) S of Ivimka Camp, 7°46.2'S, 146°29.0'E, elevation approximately 40 m, bottom of cobbles, mud, and sand with tree roots and logs, water clear and slow-flowing through open forest, pool with depth to 3.0 m, spear (G. Allen, 17 November 1996).

Station 9 - Avi Avi River, westernmost channel about 5.3 km (straight-line distance) SW of Ivimka Camp, 7°46.3'S, 146°27.9'E, elevation approximately 35 m, bottom mainly mud, and sand with tree roots and logs, water moderately turbid and slow-flowing through open forest, small side pool (3 x 10 m) with depth to 2.0 m, 1.0 kg rotenone (G. Allen and Tami [from Tekadu], 18 November 1996).

Station 10 - Small (2-3 m wide) tributary of westernmost channel of Avi Avi River, about 4.8 km (straight-line distance) SW of Ivimka Camp, 7°45.8'S, 146°28.2'E, elevation approximately 40 m, bottom of cobbles, mud, and sand, water clear and slow-flowing through open and closed-canopy forest, pools to 1 m depth with shallow sections between, 1.0 kg rotenone (G. Allen and Tami [from Tekadu], 18 November 1996).

Station 11 - Sapoi River, about 1 km upstream from "junction" with Avi Avi River, 7°43.8'S, 146°29.9'E, elevation approximately 200 m, bottom of cobbles and boulders, water clear with fast flow and slow side eddies, through open forest, depth to 2.0 m, 1.5 kg rotenone (G. Allen, 19 November 1996).

Station 12 - Small (1-2 m wide) creek, about 1 km SE of Ivimka Camp on Kakoro Track, approximately 7°44.4'S, 146°30.5'E, elevation approximately 120 m, mainly mud and sand bottom with roots and log debris, water clear and slow-flowing through closed-canopy rainforest; mainly shallow with pools to 0.7 m deep, 1 kg rotenone (G. Allen and UPNG students, 20 November 1996).

Overview of the New Guinean
Freshwater Fish Fauna
Although the marine fishes of New Guinea are reasonably well documented, there has been relatively little work on freshwater fishes. Dutch nat-

uralists made several expeditions to the interior of what is now Irian Jaya between 1903 and 1920. The majority of the fish collections were summarized by Weber (1913). Up until about 1950 the known fauna stood at approximately 150 species. Largely through the efforts of several recent workers including Ian Munro, Tyson Roberts, and myself, the total is now about 340 species. However, there still remain vast unsurveyed areas throughout New Guinea. The final species tally will perhaps approach 400 species or about twice the number known from nearby Australia.

The New Guinean fauna is very similar to that of northern Australia. In fact, there are about 35 species shared by these areas. This faunal link is not surprising considering that southern New Guinea is part of the Australian continental plate and the two areas were formerly connected by a land bridge as recently as 6000-8000 years ago.

Unlike other tropical areas such as Africa and South America, which are inhabited by numerous fishes which evolved entirely in freshwater, the species found in New Guinea and Australia, with only a couple of exceptions, are "secondary" freshwater forms. They were derived from marine fishes, which entered fresh water in relatively recent geologic times.

The freshwater fauna of New Guinea and northern Australia consists of relatively few families (about 40). The fauna is dominated by atherinoids (rainbowfishes, blue-eyes, and hardyheads), plotosid (eel-tailed) and ariid (fork-tailed) catfishes, terapontids (grunters) and gobioid fishes (gobies and gudgeons). Collectively, the eight families in these groups account for about two-thirds of New Guinea's entire freshwater fauna.

There are no previous fish collections from the Lakekamu Basin.

Results and Discussion

General
The total fauna of the Upper Lakekamu Basin reported herein consists of 23 species belonging to 18 genera and 14 families. An annotated list of species is presented in Appendix 10 and a summary list in Appendix 11. Most of the fishes appearing in the lists were discussed and illustrated by

CONSERVATION INTERNATIONAL

Allen (1991). The collections from the present survey includes at least two new goby species (Allen 1997). In addition, a species of rainbowfish (*Melanotaenia*) may also prove to be undescribed.

Faunal composition of the Upper Lakekamu Basin
The freshwater fauna of the Upper Lakekamu Basin is similar to other parts of New Guinea and northern Australia in that it is dominated by zrelatively few families. As elsewhere in this region, groups such as catfishes, rainbowfishes, grunters, gobies, and gudgeons are very common. Collectively, these groups account for 60 percent of the fauna.

Species Composition: Altitudinal Effects and Upstream Attenuation of Species
There is a general attenuation of fish diversity as one proceeds upstream in the Lakekamu Basin (Figure 1). This phenomenon is typical of other rivers in New Guinea and regions whose fauna is derived mainly from marine groups. It is also indicative of the geological youth of most of inland New Guinea. Fishes have generally had insufficient time to evolve in the interior. There are no native fishes above an elevation of about 2000 meters. This part of New Guinea has been hostile to fish evolution in recent geological times. It has been an area of tumultuous mountain building. In addition, concurrent volcanism and glaciation occurred over much of the area as recently as 300,000 years ago.

The survey site, situated over approximately 8 km of the Sapoi River, afforded an excellent opportunity to study the upstream penetration capabilities of the various species. This section of the river undergoes a very noticeable transition from a slow, braided meandering lowland (about 35 m elevation) stream with deep pools and relatively slow flow to a torrential mountain (about 200 m elevation) stream characterized by uninterrupted rapids and cascades. Approximately 19 species were present in the lowlands 6 km downstream from Ivimka Camp compared to only six species above the first set of formidable cascades, situated about 2 km upstream from Ivimka Camp (Table 1).

The observed upstream species attenuation was no doubt caused by the disappearance of preferred habitats and increasing current velocity. For example, a number of the lowland fishes such as the fork-tailed catfish (*Arius*), garfish (*Zenarchopterus*), and freshwater cardinalfish (*Glossamia*), are dependent on the presence of deep pools with minimal current. Two of the most common residents of the river, the rainbowfish *Melanotaenia goldiei* and the grunter *Hephaestus trimaculatus*, live in a variety of habitats, often in sections where currents are very swift. But both of these species are unable to pass above the first major cascades, situated about 2 km upstream from Ivimka Camp.

Figure 1. Upstream attenuation of Lakekamu fishes. Generally, species numbers attenuate as one proceeds upstream through the survey area. The zero kilometer mark on this graph represents the farthest point sampled downstream, which was about 6 km from Ivimka Camp.

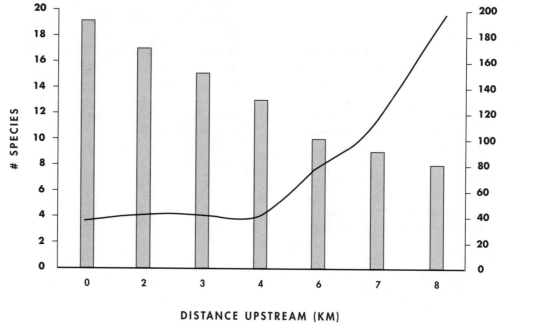

CONSERVATION INTERNATIONAL

Rapid Assessment Program

Table 1. Upstream penetration of Lakekamu fishes. The distance column on the left indicates the approximate upstream penetration from the Lakekamu River mouth. The right column refers to the distance upstream (+) or downstream (-) from Ivimka Camp. Values shown with an asterisk (*) indicate that the upper limit of penetration was not detected during the survey and probably lies well upstream, in mountainous terrain.

SPECIES	DISTANCE FROM SEA (KM)	DISTANCE FROM IVIMKA	HABITAT PREFERENCE
Anguilla bicolor	100+	+2*	fast rocky streams
Arius leptaspis	91	-4.5	deep pools, slow current
Neosilurus ater	90	-5.5	deep pools, slow current
Neosilurus brevidorsalis	100+	+2*	pools, slow-fast current
Zenarchopterus novaeguineae	90	-5.5	pools, slow-current
Melanotaenia goldiei	97	+2	rocky streams, slow or fast
Melanotaenia rubrostriatus	91	-4.5	deep pools, slow current
Melanotaenia sp.	93	-1.5	pools in forest streams
Craterocephalus randi	95	+0.5	sunlit pools, slow or medium
Hephaestus trimaculatus	97	+2	rocky pools
Kuhlia marginata	100+	+2*	pools at base of rapids
Glossamia sandei	93	-1.5	log jam pools, slow current
Oreochromis mossambica	92	-3	pools, slow current
Cestraeus goldiei	100+	+2*	pools at base of rapids
Crenomugil heterocheilus	92	-3	pools, shallows, slow current
Awaous acritosus	95	+0.5	gravel bars, slow or medium
Glossogobius celebius	90+	Avi Avi	sand or mud, slow to fast
Glossogobius sp.	100+	+2*	cobbles, slow to fast
Lentipes watsoni	100+	+2*	rocky, very fast current
Mogurnda pulchra	95	+0.5	mud, slow streams or swamp
Oxyeleotris fimbriata	100+	+2*	mud or rock, slow to fast
Oxyeleotris gyrinoides	95	-0.3	mud bottom of slow streams
Synaptura villosa	95	+0.5	sand or mud, slow to med.

Table 2. Behavioral modes and feeding relationships: the main activity modes of Lakekamu fishes. All the fish species (23) could be assigned to one of six behavioral foraging modes.

BEHAVIORAL MODE/FEEDING RELATIONSHIP	PERCENT (%) OF SPECIES
Diurnal benthic	52.2
Diurnal midwater	26.1
Nocturnal benthic	8.7
Cryptic	4.3
Surface swimmer	4.3
Benthic grazing schools	4.3

Well over one-half of the fauna is composed of species that live on or near the bottom. The most prominent in this respect are gobies and some gudgeons, which rest directly on the substratum or catfishes and grunters, which actively swim over it. The remaining fauna is composed of fishes that either actively swim or hover well above the bottom. This category includes small hovering gudgeons and more active forms such as rainbowfishes, hardyheads, and halfbeaks. The latter groups often school in large numbers close to the surface.

There are three main feeding types of Lakekamu fishes (Table 2). The majority of species are carnivorous (52.2%), which is typical for both tropical streams and coral reef systems. Aquatic and terrestrial insects and various aquatic larval insects feature prominently in the diet of numerous small fishes, particularly rainbows. The remaining 48 percent of the fauna consists of fishes that are either purely herbivorous (17.4%) or omnivorous (30.4%). Of the herbivores, the mullets are particularly prominent. *Crenomugil heterocheilus* often forms schools in shallow sunlit sections of the river where it grazes green filamentous algae from rocky substrates. The Three-spot Grunter, *Hephaestus trimaculatus* is one of the most prominent omnivores. It consumes a variety of invertebrates including prawns and insects, but it also ingests a significant amount of algae.

Zoogeographic Affinities of the Upper Lakekamu Basin Fish Fauna

Zoogeographic categories of Lakekamu fishes are summarized in Table 3. The majority of species belong to the overall faunal community of the northern Australia-New Guinea region. Although most families and genera are consistently present across the region, the species composition varies greatly according to locality.

The majority or 70 percent of Lakekamu fishes are distributed widely either across the southern portion of New Guinea or the combined northern Australia-southern New Guinea region. Most of the remaining species are relatively widespread in either the Indonesian-Melanesian Archipelago or beyond. These are forms that presumably possess a marine larval stage, but there is still much to be learned about the life histories of all freshwater fishes in New Guinea.

Table 3. Zoogeographic analysis of Lakekamu fishes. Species occupy known ranges that are readily attributable to six biogeographic regions. Note that one of these is an exotic, introduced species.

ZOOGEOGRAPHIC REGION	NUMBER OF SPECIES AND (% OF FAUNA)
South New Guinea	11 (48%)
South New Guinea and north Australia	5 (22%)
Western Pacific	2 (9%)
Indo-West Pacific	2 (9%)
Indo-Australian Archipelago	2 (9%)
African introduction	1 (4%)

Regional Endemism

The Lakekamu drainage lies within an area that has been poorly sampled for fishes. It is situated between the Great Southern and Eastern Papua zoogeographic provinces as defined by Allen (1991) and contains faunal elements from both areas. The overall fauna seems strongly linked to the Great Southern Province, which includes most of the southern half of central New Guinea. However, the presence of the grunter *Hephaestus trimaculatus* and gudgeon *Mogurnda pulchra* indicate a relationship to the Eastern Papua Province.

There are probably no endemic fishes in the Lakekamu drainage system, although at present the new gobiid fish, *Lentipes watsoni*, is known only from the Sapoi River. Future collections will no doubt expand the range to include adjacent drainages. In addition, the status of the undescribed *Glossogobius* and possible new *Melanotaenia* are currently under investigation by the author.

Introduced Fishes

Freshwater systems of the New Guinea-Australia region have relatively few species compared to their counterparts in South-east Asia, Africa, and South America. Therefore, the introduction of exotic fishes is more likely to have immediate detrimental consequences. Introductions are likely to compete for food and living space, and in some instances they may feed on the young or adults of native species. There is absolutely no justification for the introduction of any fish species anywhere

in New Guinea. Unfortunately, about 15 species have already been introduced, mainly for fisheries purposes.

A single introduced species, the African Tilapia (*Orechromis mossambica*) was detected in the present survey. Only three individuals were seen, three juveniles (about 5-8 cm SL) were seen of which two were collected. Therefore, it is apparent that the species has not gained a foothold in the Upper Lakekamu Basin. Likely, it is more common and has a breeding population farther downstream in the meandering lowland section of the Lakekamu River. Tilapia was widely introduced in New Guinea as a food source by the Department of Stock, Agriculture, and Fisheries, beginning in 1954 (Glucksman *et al.* 1976).

River Damming - A Unique Method of Fishing

The school teacher at Tekadu, Joe Abraham, and another resident of the village gave a fascinating account of a method occasionally used to catch fishes in the Avi Avi River. Apparently, about once every decade, there is a prolonged drought and the river level is significantly lowered. The residents of Tekadu, Kakoro, and Nukeva join together to completely block the Avi Avi River, diverting the flow to the nearby Sapoi River. This is done about 2 km upstream from Ivimka Camp, at the point where the two rivers are less than 100 meters apart, and are separated by a low ridge of sand, gravel, and rock. The men dam the river by constructing a strong, reinforced "fence" of bush poles and thatching. Once this task, which usually takes about 3 days, is completed, the women add the final touch by filling in all the spaces with leaves. Once the flow is diverted into the Sapoi, the downstream portion of the Avi Avi ceases to flow. Fishes are stranded in the pools and are easily speared by the villagers, who live in specially constructed fishing camps. The river was dammed in this manner in 1977 and 1987. In 1987, the fishing lasted for nearly 3 months and fishes were regularly carried to Kakoro and Tekadu, then exported to Wau.

Conclusions and Recommendations
Fishes are an important part of the Lakekamu biota. They should form an integral part of any conservation-based programs or strategies in the area for the following reasons. The Upper Lakekamu Basin, the Sapoi River in particular, offers a pristine habitat, which except for the rare occurrence of introduced Tilapia, has a rich assemblage of native fishes. Such pristine freshwater environments are becoming increasingly scarce in PNG; the recent growth in logging will negatively impact rivers in many parts of the country. The fish fauna of the area surveyed is unique: the Sapoi River and adjacent Avi Avi River is the only know habitat of the Clinging Goby, *Lentipes watsoni*. Additionally, the Upper Lakekamu Basin is home for another undescribed goby (*Glossogobius*) and a possible new rainbowfish (*Melanotaenia*). There is a possibility that these fishes may be confined to the Lakekamu Basin and adjacent drainages. Lastly, the fish of the Lakekamu Basin are an important dietary resource for the people living in the Basin. Without fish, many people's diet would be low in protein, potentially leading to deleterious health effects. The maintenance of good water quality, coupled with wise fishing practices would ensure this important resource is available to future generations in the Basin.

It would be worthwhile to expand our knowledge of the Lakekamu drainage system by expanding the coverage of future fish surveys in both downstream and upstream directions. It would be particularly interesting to complete the investigation of upstream penetration by the six species occurring at the upper limit of the present study. Downstream, the fauna could be expected to increase dramatically. I would estimate an overall fish total of 60-70 species, with most additions coming from the lowermost sections of the river in mangrove and Nipa-palm habitat. Appendix 12 compares the Upper Lakekamu Basin fauna with that of other river systems in New Guinea.

The Upper Lakekamu Basin, the Sapoi River in particular, offers a pristine habitat ... such pristine freshwater environments are becoming increasingly scarce in PNG;

HERPETOFAUNA
(Allen Allison, David Bickford, Stephen Richards and Geordie Torr)

Chapter Summary

• Herpetofauna of the Lakekamu Basin was sampled by general collecting, leaf litter plots and visual encounter surveys.
• Seventy-four species of amphibians and reptiles (amounting to roughly 14.2% of the known herpetofauna of PNG) were recorded.
• Eleven species of frogs and seven species of reptiles may be new to science.
• Frog diversity and density were high relative to other southern lowland sites in PNG.
• Frog populations were healthy and vigorous, making Lakekamu an important site for ongoing monitoring given the widespread decline of amphibians worldwide.
• Lakekamu now has one of the best-known herpetofaunal assemblages in PNG.

Introduction

A total of 521 currently recognized species of amphibians and reptiles have been described from PNG, of which 200 (~40%) are endemic (Allison, 1993, 1996 and unpublished updates). However, new species continue to be discovered, and many groups require taxonomic revision; the actual species count for PNG is almost certainly closer to 700 (Allison 1993) and the level of endemism probably exceeds 50%.

The herpetofauna of PNG comprises about 5% of the world's known reptile and amphibian species, highlighting the importance of this country to global biodiversity. The probable occurrence of many undescribed species in remote regions of the country, and our poor knowledge of the classification and distribution of most species, emphasizes the importance of further research on the herpetofauna. In particular, information on the systematics and biogeography of PNG's herpetofauna is crucial to efforts to conserve this unique assemblage.

Reptiles and amphibians are important components of New Guinea ecosystems. They include primary consumers (*e.g.*, tadpoles; herbivorous lizards) and large terrestrial and aquatic predators (*e.g.*, pythons and soft-shelled turtles). Indeed, the largest terrestrial predators in New Guinea are varanid lizards and pythons. In addition, many reptile and frog species occur in relatively high densities and are an important food source for other organisms and for the local people.

Because frogs have a permeable skin and often have an aquatic life stage, they are especially susceptible to minor changes in temperature, humidity, water chemistry, and environmental toxins. Recent declines in amphibian populations on a global scale have alerted scientists to possible environmental changes that could eventually also impact other taxa. Careful monitoring of amphibian populations may provide important insight to environmental change, and enhance the potential for corrective action. Such monitoring requires baseline data from many localities, including remote locations with rich amphibian populations, as are found in PNG. Inasmuch as many New Guinea frogs are in the same genera as, and ecologically similar to, species that have suffered declines in northern Queensland, Australia (Richards, McDonald and Alford 1993), information from PNG is potentially important to understanding environmental change at both a regional and global level.

Our efforts during the RAP were concentrated on inventorying and quantifying the populations of amphibians and reptiles around the Ivimka camp. This site is included in the Lakekamu Basin/Chapman Range area listed as important in the PNG Conservation Needs Assessment (Beehler, 1993) and is within the area cited by Allison (1993) as poorly known herpetologically.

Small collections from nearby areas, and extensive collections from Brown River near Port Moresby, and Bereina, 50 km east of the Lakekamu River, have indicated that the herpetofauna of the Lakekamu Basin may be distinct from that of south coast lowlands to the immediate east. Our objectives were to document the distribution, relative abundance and diversity of the herpetofauna, and to learn more about the natural history of the organisms, especially in terms of reproduction and life history traits (frog calls, egg laying, mating behaviors, tadpole developmental series, and clutch size). An important additional objective was to gather quantitative data on amphibian populations in order to add to a small but growing dataset on the status of amphibian populations around the world.

Methods

The Rapid Assessment was conducted from 24 October to 5 December 1996. Two of us (S. Richards and G. Torr) were engaged in other herpetological research at the site that yielded data incorporated into this report. Additionally, people from Tekadu village assisted with general collecting, bringing the RAP team many interesting specimens. Although non-quantitative general collecting has been historically the most appropriate way of quickly assessing the herpetofaunal assemblage at a given site, one of us (D. Bickford) has initiated long term monitoring of frog populations at different sites in PNG and continued these standard methods for comparative purposes and standardization during the RAP (see Heyer *et al.* 1994).

Following a basic general collecting regime, we sampled small pools and streams for tadpoles, rearing the tadpoles which could not be immediately identified. We searched microhabitats that are known to be especially productive for herpetofauna, *e.g.*, riparian associations, epiphytes, under fallen logs, loose bark, tree buttresses, rock crevices, *etc.*

Twenty-five nocturnal visual encounter survey (VES) transects were conducted (D. Bickford). Each VES entailed walking a transect for one hour, systematically searching with a headlamp

Our poor knowledge of the classification and distribution of most species, emphasizes the importance of further research on the herpetofauna.

for frogs. All individuals seen were identified, measured (snout-vent length [SVL]), and sexed. The substrate where the animal was located was recorded and individuals were either released or collected. Transects traversed a variety of micro-habitats in order to thoroughly inventory the community.

Forty randomly located leaf litter quadrats, each 5m x 5m, were sampled in ten mornings (4 plots per morning). Three field assistants (from a pool of eleven UPNG students and three local assistants) and D. Bickford each searched one side of each plot, clearing leaves in concentric perimeter sweeps until the plot was completely cleared. All individuals encountered were caught and measured as in the VES, then either collected as vouchers or released upon completion of the quadrat. Additional quadrat habitat measurements included: number of logs, number and sizes of trees, moisture, canopy cover, temperature, leaf litter depth; rockiness, vegetation type, and ground slope.

General collecting of voucher specimens (excluding protected species) was undertaken to yield a well-determined series from the site in order to facilitate future research and reduce the necessity of future collecting. We euthanized all specimens by submersion in chlorotone (for amphibians and small reptiles), or with lethal injection of pentobarbital in the cardiac muscle (for large reptiles). Tissues (cardiac muscle, liver, and oocytes, if present) were prepared immediately by immersion in 95% ethanol or 20% DMSO for future molecular systematic analysis. We prepared specimens for collections and morphological work by either fixing in 10% formalin solution, and then storing in 70% ethanol, or by both fixing and storing in 70% ethanol.

Results

Seventy-four species of amphibians and reptiles were recorded during the RAP (Appendices 13 and 14). Thirty species of frogs and forty-one species of reptiles were collected as specimens; three reptile species were positively identified but not collected.

Species/effort curve for the leaf litter plots appears to level off at five species after just six sampling periods or twenty-four plots (Appendix 1H1). Twenty-one individuals were found in the 1000 m^2 sampled, or roughly one frog per 48 m^2 (Table 1). *Hylophorbus* sp. was the dominant species, representing more than half of the frogs in the plots.

Sixteen species were encountered on the VES transects (Appendix 1D). Eighty frogs were found on VES, for an average encounter rate of 3.2 frogs/hour (Table 2). Again, *Hylophorbus* was one of the two most common species found. The high number of *Rana grisea* were mostly juveniles dispersing away from the streams. Ten of the 16 species were encountered only 1-2 times.

Discussion

The leaf litter frog fauna at Ivimka is relatively species rich and dense compared to another lowland site in the Gulf Province (So'obo Camp, Crater Mountain Wildlife Management Area, D. Bickford unpublished data). At So'obo only one individual of each of two species was found with the same number of leaf litter plots as surveyed at Ivimka (Table 1). This large difference could be partly due to seasonal effects, but it suggests that Ivimka has a comparatively rich leaf litter frog community.

The frog assemblage at Ivimka is comprised of a few common frog species and many rare species, a pattern typical of lowland rainforest communities. Fifty-eight percent of the frogs found on plots or VES were of two species: *Hylophorbus* sp., and *Rana grisea* (Tables 1 and 2).

After 15 days of fieldwork, the quantitative methods yielded data that can be statistically compared with other sites or subsequent censuses but they are not exhaustive for all frogs. Five species of frog were found using the leaf litter plot methodology and the rarest terrestrial or fossorial frogs were not discovered. Moreover, the VES transects sampled only sixteen species or 53% of the frog species encountered on the survey. This does not invalidate these methods, however, as they revealed several species which were otherwise not found. Given the species richness and diversity of habitats occupied by tropical frog assemblages, it is desirable to utilize a variety of sampling techniques, as field time allows, to thoroughly inventory rainforest frogs.

Our general life history observations are incorporated into species accounts (Appendix 15). Eleven species of frogs did not reliably match any of the currently recognized species, and very likely represent new taxa. Likewise seven species of reptiles may represent new taxa. In some cases the genera involved (*e.g.*, the microhylid genus *Oreophryne*) are too poorly known to make it possible to readily recognize new species. In other cases (*e.g.*, the scincid lizard genus *Prasinohaema*), one of our specimens definitely represents a new taxon. Many lizard groups have not been comprehensively treated since de Rooij's (1915) monumental work.

A comprehensive zoogeographic analysis of the Lakekamu herpetofauna is beyond the scope of this paper. However, our collections document significant eastern range extensions of at least three species. These include a frog *Callulops slateri*, and lizards *Emoia physicina, E. tropidolepis*, and possibly *Hypsilurus* cf. *auritus, Cyrtodactylus* cf. *mimikanus*. In addition species in the

microhylid frog genus *Xenobatrachus* are generally restricted to western PNG and Irian Jaya. Areas to the east of Lakekamu have been relatively well collected. The absence of the above mentioned species from that region suggests that the Lakekamu Basin may represent a zone of faunal disjunction, perhaps reflecting the geological past (see Davies et al. 1996). The main affinities of the herpetofauna seem to be with elements to the west.

The quantitative data produced on the RAP provide a baseline for future monitoring programs. Local landowners can be trained as field workers to conduct long-term monitoring, as has been done at other sites in PNG. The establishment of an on-going amphibian monitoring program at Lakekamu may be useful for:

1) evaluating the effectiveness of some aspects of the Lakekamu Basin conservation initiative;

2) monitoring frog populations as part of the international effort to assess global amphibian population declines;

3) producing additional data on the ecology of PNG's rainforest frogs; and

4) involving local landowners and PNG students through direct participation in research at the Ivimka Research Station.

Table 1. Frog species encountered in 5 x 5 meter leaf litter plots at Ivimka camp, Lakekamu Basin 120 m - 180 m asl. See text for details of methods.

SPECIES	NUMBER OF INDIVIDUALS	DENSITY
Hylophorbus sp.	11	1 per 91 m^2
Copiula sp.	7	1 per 143 m^2
Lechriodus melanopyga	1	1 per 1000 m^2
Rana grisea	1	1 per 1000 m^2
Xenobatrachus sp.	1	1 per 1000 m^2

Table 2. Frog species encountered in nocturnal visual encounter survey (VES) transects at Ivimka camp, Lakekamu Basin 120 m - 180 m asl. See text for details of methods.

TAXA	NUMBER OF INDIVIDUALS	ENCOUNTER RATE
Rana grisea	24	0.96/hour
Hylophorbus sp.	24	0.96/hour
Lechriodus melanopyga	6	0.24/hour
Litoria pygmaea	4	0.16/hour
Litoria dorsalis	3	0.12/hour
Callulops sp.	3	0.12/hour
Mantophryne lateralis	2	0.08/hour
Cophixalus cheesmani	2	0.08/hour
Oreophryne sp 1	2	0.08/hour
Rana arfaki	2	0.08/hour
Callulops slateri	2	0.08/hour
Litoria genimaculata	2	0.08/hour
Litoria sp. 3	1	0.04/hour
Copiula sp.	1	0.04/hour
Xenobatrachus sp.	1	0.04/hour
Litoria sp. 2	1	0.04/hour

BIRDS
(Andrew L. Mack and Paul Igag)

Chapter Summary

• One hundred-ninety species of birds (= 31% of PNG's resident avifauna) have been recorded in the Lakekamu Basin, of which 132 were recorded on the RAP survey, making the Basin the richest and among the best-known sites in all of New Guinea.

• Four threatened and six near-threatened species (IUCN 1996) have been recorded from the Basin.

• Seven species were recorded for the first time in the Lakekamu Basin, indicating the basic inventory of the avifauna is nearly complete. Future ornithological work should emphasize more-detailed ecological study.

• Point counts yielded data on 68 species. Fifty-one species and 467 individuals were captured and banded. These data provide a foundation for ongoing, long-term ecological studies at the Ivimka Research Station.

• Comparison of RAP survey data with data collected in other parts of the Basin indicate that bird populations are not uniformly distributed across the Basin. Conservation of the full spectrum of the Basin's 190 species will require conservation of a large expanse of Basin forest. Undoubtedly, this conclusion also applies for other taxa that have not been surveyed as thoroughly as birds.

Introduction

Up until 1979 no ornithologists had visited the Lakekamu Basin and few biologists had visited the extensive wet forests along the southern flank of the Central Ranges of PNG. Within the past twenty years exploration and research in the southern lowlands and mid-elevations has grown (*e.g.*, Bell 1982, Bell 1984, Hyndman and

Menzies 1990, Mack and Wright 1996). Generally, the avifauna of the wet forests of southern PNG west of Port Moresby constitute one of the major faunal zones of New Guinea. This zone is marked by high species richness in most taxa, but only among birds is this diversity well documented and distributions broadly understood. For most other taxa we have only the sparsest knowledge of systematics and biogeography and even less about ecology.

Within this zone, the Lakekamu Basin is an outstanding tract of lowland forest due to its extent and relative freedom from anthropogenic modifications. In 1979 Dr. Bruce Beehler made the first of many visits to the Lakekamu Basin where he and his associates have conducted extensive ornithological and ecological studies (Beehler *et al.* 1994, Beehler *et al.* 1995). The expansive undisturbed forest, the high species richness and the pioneering studies of Beehler *et al.* make the Lakekamu Basin a high priority area for conservation (DEC 1993) and an ideal location for much-needed research. The threats to the Basin posed by growing interests in timber, goldmining, plantation forestry, and an expanding human population add urgency to the need for conservation in the region.

Methods

Birds were surveyed using three methods: general searching, point counts, and mist-netting. General searching provides an inventory of the birds evident in the area. Point counts provide crude quantitative data on abundance of singing birds. Mist-netting provides crude quantitative baseline data on understory birds and their phenology. These data are compared with netting data from other parts of the Basin (Beehler *et al.* 1995).

General Observations
All species observed during the survey period (15 October-15 November) and subsequent training course (16 November-12 December) were recorded, including those noted on point counts or netted. Bird songs were tape recorded whenever possible. Mixed flocks were followed and efforts were made to observe birds in a variety habitats.

Observations by other observers in the RAP team were included in the final species tally.

Point Counts
Thirty-three point counts were made at 200 m intervals along the trail system. On each count, lasting ten minutes, all birds heard or seen were recorded.

Mist-Netting
Mist nets were set on the inner loop trail and the first three 1-km transects off the Bulldog Road in the flat alluvial forest and a line of nets was run on the ridge trail above the research station for 3 days(see map page 22). Each line of nets (consisting of 39 nets: 9, 12 and 15 m lengths for a total of 402 meters of net) was run for at least three days as weather permitted. Birds were banded using Australian Bird and Bat Banding Scheme bands and the data added to the Australian computerized database. Standard data (weight, wing chord, culmen, molt score, fat, and presence of brood patch) were recorded. Some voucher specimens were collected (Appendix 16).

Results

General Observation
A total of 132 species were observed during the period of the RAP survey and RAP training course (Appendix 16). The species accumulation curve reached an asymptote suggesting continued sampling would not reveal many new species (Appendix 1E).

Seven species were recorded for the first time in the Basin: *Halycyon nigrocyanea, Hirundapus caudacutus, Pachycephala hyperythra, Sericornis spilodera, Myzomela nigrita, Erythrura papuana,* and *Cicinnurus magnificus,* re-emphasizing that this is the most species-rich site known in New Guinea.

The Basin is home to several bird species of special conservation importance. *Casuarius casuarius, Harpyhaliaetus novaeguineae, Goura scheepmakeri* and *Psittrichas fulgidus* are listed as Threatened species on the IUCN Red List (IUCN 1996). Additionally, six Lakekamu Basin species are listed as near-threatened (IUCN 1996):

The Lakekamu Basin is an outstanding tract of lowland forest due to its extent and relative freedom from anthropogenic modifications.

Zonerodius heliosylus, Aquila gurneyi,
Henicophaps albifrons, Trugon terrestris,
Probosiger aterrimus, and *Aplonis mystacea.*

Point Counts

A total of 932 individuals were recorded during
the 330 minutes of point count observation ($\bar{x}$ =
28.3 individuals per count SD=8.51, range= 12-
48). Sixty-eight species were recorded during
point counts ($\bar{x}$ = 15.5 species per count, SD=
3.20, range= 9-23).

There was close agreement between the species
recorded on the most point counts (frequency)
and the species most numerous on point counts
(abundance). These were those species that are
widespread and vocally conspicuous at this time
of year: *Trichoglossus haematodus* (and *T.
goldiei*), *Pitohui ferrugineus, Dicrurus hottentot-
tus, Lorius lory, Xanthotis flaviventer, Ptilorrhoa
caerulescens, Paradisaea raggiana, Ptilinopus
pulchellus, Philemon buceroides, Mino dumontii,
Colluricincla megarhyncha,* and *Talegalla fus-
cirostris* (in descending order of frequency and
abundance).

Point counts rely upon observer capability and
seasonality of vocalizations (Bibby *et al.* 1992).
The counts during the RAP plateaued between 60
and 70 species after about 20 (200 minutes) point
counts (Appendix 1F). Thus, point counts quickly
gave a quantitative assessment of vocal activity
of a subset of the birds but did not yield a complete list of the resident avifauna. Point counts
do, however, enable ornithologists to determine
seasonal cycles of vocal activity— often a correlate to breeding activity. It will be enlightening
for future ornithological researchers in the Basin
to conduct point counts at different times of year.

Mist-Netting

Mist nets were run for a total of 77,414 net meter
hours. The species accumulation curve (Appendix
1G) gently climbed throughout the survey period,
as is typical for mist netting studies in tropical
forests (Mack and Wright 1996). We recorded
754 captures of which 135 were recaptures; 467
individual birds were banded.

Fifty-one species were netted with a range of
one to 73 individuals per species. The number of
rare versus abundant species was not substantially
different between Lakekamu and Crater Mountain
(a hill forest to submontane site) but the Crater
site did have more abundant birds (Table 1). In
other words, the most common understory birds
at Lakekamu were not as abundant as at the higher site, but overall species richness was higher in
the lowland forest. This is reflected in the capture
rate, standardized to sampling effort between the
two sites: on the RAP roughly 8 birds per 1000
meter hours versus 13 birds per 1000 meter hours
at Crater.

The abundant understory birds at Crater were
not the most common at IRS (Crater *vs.* IRS captures per 100 net meter hours): *Toxorhamphus
poliopterus* (1.94 *vs.* 0.36), *Oedistoma iliolophus*
(1.19 *vs.* 0.03), *Melilestes megarhynchus* (0.75 *vs.*
0.16), *Sericornis spilodera* (0.46 *vs.* 0.01), *and
Pachycephalopsis poliosoma* (0.44 *vs.* 0.0).
Whereas the common understory birds at IRS
were generally uncommon at Crater (IRS *vs.*
Crater captures per 100 net meter hours):
Monarcha guttula (0.94 *vs.* 0.21), *Monarcha
manadensis* (0.57 *vs.* 0.0), *Micropsitta pusio* (0.44
vs. 0.0), *and Ceyx lepidus* (0.44 *vs.* 0.03).

Table 1. Comparison of mist-netting data from the lowland Lakekamu site and a nearby site at 1000 m elevation.

NUMBER OF INDIVIDUALS	RAP SPECIES (individuals)	LAKEKAMU*SPECIES (individuals)	CRATER MOUNTAIN**SPECIES (individuals)
1-2 (rare)	16 (19)	29 (39)	19 (22)
3-19 (common)	22 (173)	21 (161)	30 (225)
> 20 (abundant)	12 (421)	20 (945)	15 (1207)

* Includes data from Beehler *et al.* 1995 and RAP data.

** Data from Mack and Wright 1996.

CONSERVATION INTERNATIONAL

A substantial difference in avifauna would be expected between Crater and IRS due to their different elevations and rainfall regimes (Table 1). However, it is surprising to note the apparent differences in understory avifaunas among sites within the continuous alluvial forest in the Lakekamu Basin (Table B2).

Table 2. Most commonly-netted species at four sites in the Lakekamu Basin. Percentage of captures are compared to show relative abundance of understory bird species varies within the Basin.

SPECIES	IVIMKA (%)[1]	KAKORO (%)[2]	SI R. (%)[3]	NAGORE R. (%)[4]
Monarcha guttula	12	9	4	4
Melanocharis nigra	9	2	1	5
Monarcha manadensis	7	3	4	8
Meliphaga aruensis	6	7	3	3
Micropsitta pusio	5	3	0	1
Colluricincla megarhyncha	5	17	4	3
Ptilorrhoa caerulescens	5	4	11	4
Toxorhamphus novaeguineae	5	7	8	2
Ceyx lepidus	4	1	0	4
Lichenostomus obscurus	4	3	0	1
Crateroscelis murina	3	2	3	1
Rhipidura threnothorax	3	3	4	3
Gerygone chrysogaster	2	3	0	1
Tanysiptera galatea	2	0	13	11
Rhipidura rufidorsa	2	3	0	0
Arses telescophthalmus	2	0	0	1
Ailuroedus buccoides	2	1	3	3
Halcyon torotoro	2	0	0	0
Melilestes megarhynchus	2	2	13	3
Ptilinopus pulchellus	1	3	0	3
Pitohui ferrugineus	1	4	0	8
Ptiloris magnificus	1	3	0	1
Pomatostomus isidorei	1	1	7	3
Xanthotis flaviventer	1	3	4	3
Dicrurus hottentottus	1	1	1	4
Chalcophaps stephani	1	2	0	4
Cracticus quoyi	1	0	6	3
Poecilodryas hypoleuca	1	6	0	0

[1] N= 619 captures. 1996 RAP

[2] N= 117 captures, Beehler et al. 1995

[3] N= 71 captures, Beehler et al. 1995

[4] N= 518 captures, Beehler et al. 1995

Discussion

Point counts provide a crude relative measure of frequency. Because these counts are dependent upon observer ability and to variations in avian behavior, they are of limited utility for estimating populations of birds in New Guinean tropical rainforests. However, they do give a rough measure of avian activity, particularly among conspicuous taxa. If point counts are performed regularly over several annual cycles at the Ivimka Research Station, they will provide insight to the phenological cycles of birds. The data presented here are hoped to act as a foundation and stimulate continued research on avian phenology at the site. The vast majority of birds recorded on point counts (>90%) were only heard, emphasizing the importance of having trained biologists to undertake avian censuses in PNG (Beehler *et al.* 1995).

The species accumulation curve (Appendix 1E) levels off, suggesting most of the species evident at this time were recorded. Beehler *et al.* (1995) recorded 108-159 species during their various censuses in the Basin; thus, this site seems typical of the Basin in terms of avian species richness. The site does not have two habitat types that would increase overall species richness due to lack of the habitat specialists that occupy them (large rivers for aquatic birds or large [anthropogenic] disturbances for grassland and second growth species). Despite the intensive efforts during the survey, only seven species were added to the Basin species list. Thus, it appears we are approaching a reasonable avian inventory in the Basin. However, we know very little of the ecology of these species. Clearly, they are not uniformly distributed within the Basin. Little is known of the relationships between each species and the complicated habitat and micro-habitat mosaics that comprise the Basin. Because birds are highly mobile and conspicuous, they might be observed in many parts of the habitat mosaic, but they could have specific habitat requirements that are essential but unapparent without detailed study. Ornithology in Lakekamu is ready to make the transition from basic inventories to more-detailed ecological assessments.

Mist-netting captures indicate a species rich understory bird community with many relatively uncommon species. Comparison with netting data from a higher location (Table 1) further suggests that the lowland avifauna is more species rich, with more rare species whereas the higher site had a few understory species that were more strongly dominant than the most numerous understory species at Lakekamu. Comparisons among sites within the Basin (Table 2) show that even among the common understory birds there are substantial differences between sites in the same continuous forest. For example, the fifth most commonly-netted bird at Ivimka, *Micropsitta pusio*, was not netted at Si River and the most-commonly netted bird at Si and Nagore Rivers, *Tanysiptera galatea*, was rarely netted at Ivimka and never at Kakoro (Table 2). It is probable these apparent differences in bird populations reflect subtle differences in habitat. Conserving the full range of species found in the Basin will require a large conservation project that encompasses adequate area of the full array of habitat mosaic within the Basin.

Several globally threatened species have intact populations within the Lakekamu Basin. Cassowaries, Goura pigeons and Harpy Eagles are often heavily hunted in much of PNG and can be expected to disappear when logging or development occurs in the lowlands. The large, undisturbed forest of the Lakekamu Basin presents an exceptional opportunity to conserve these stunning birds. Areas should be set aside, preferably including the area surrounding the IRS, where local hunters agree not to hunt. These zones would act as population source areas and enhance the research potential at IRS. Another threatened species, *Psittrichas fulgidus*, was also found in the Basin and conservation here could substantially benefit these unique parrots. However, it is possible (Mack and Wright 1998) that *Psittrichas* also require foothills and middle elevations for population viability. If hills and steep regions of the adjacent ranges remain intact, species such as *Psittrichas* that utilize resources outside the lowland alluvial Basin can probably be conserved also.

The data collected during mist netting lay an important foundation for future studies at the Ivimka Research Station. Data on molt and breeding condition were collected for most of the 467 birds banded. Recapture of these birds in the future will shed light on the seasonal activities of birds in the Basin, movement patterns, longevity, and other important aspects of their ecology.

Aplonis mystacea, a species of notable interest in the Basin (Beehler and Bino 1995, Safford 1996), was not observed during the survey. However, it was commonly observed during a reconnaissance trip in April the same year. Undoubtedly, the birds move around the Basin depending on the availability of their preferred fruiting trees. As recommended by Beehler *et al.*, the population in the Basin should be studied in order to learn its conservation requirements. This recommendation should be extended to the other threatened and near-threatened species listed by the IUCN that occur within the Basin.

MAMMALS
(Debra D. Wright, Solomon Balagawi, Tau Teeray Raga, and Rodney T. Ipu)

Chapter Summary

• Using nocturnal observation, traps, mist-nets and harp traps we confirmed the presence of 3 marsupial species, 12 rodent species, and 9 bat species (24 total species) in the IRS study area.
• A literature search suggested as many as 80 mammal species could occur within the Basin.
• Of the 24 mammal species recorded, The IUCN Red Data Book lists one as "threatened" (*Hipposideros muscinus*), two as "vulnerable" (*Paranyctimene raptor* and *Pipistrellus wattsi*), and one as "data deficient" (*Melomys levipes*).
• Species accumulation curves showed that a minimum 3-week effort would have been sufficient to find this diversity.
• *Rattus leucopus* and *Syconycteris australis* were the most commonly encountered species.
• A number of novel and interesting natural history observations were made.

Introduction

Currently, we know of 212 mammal species in New Guinea (Flannery 1995). Of these 59 are on the IUCN Red List 1 for Threatened Animals (IUCN 1996), and an additional 27 (six more should be included according to Flannery [1995] are on List 3 for Near Threatened (vulnerable) Animals. Another 22 species are on List 5 for Data Deficient Animals (28 more should be included according to [Flannery 1995]. Thus, 43% of New Guinea's mammals are in need of conservation measures and we need more data on another 24% which may be in need of conservation also. It appears that roughly half of the mammal species in New Guinea need help to ensure their continued existence; of these, 85% are endemic to the mainland or the immediately surrounding islands. While these staggering numbers may in part reflect the difficulty of observing and capturing mammals in New Guinea, we should not ignore the implications.

Given Lakekamu's location and elevation, 80 known mammal species could occur within the Basin. Of these 24% are considered threatened or vulnerable and an additional 10% are data deficient. Therefore, one-third of the probable mammal species within the Basin are in need of conservation or research measures.

We know very little about New Guinea mammals. Although surveys have been conducted to determine ranges, for a surprising number of species we know nothing about reproduction and can only surmise diet (Flannery 1995). New Guinea mammals are faced with the indirect threat of habitat destruction and with the direct threat of over-hunting by local people. Terrestrial and ground-denning species (*e.g.*, wallabies, some cuscus) are particularly vulnerable to hunting with dogs. Cave-roosting bats are also exceptionally vulnerable.

Virtually no mammal work has been done in the Lakekamu Basin. These data were collected to begin the task of documenting the mammalian fauna of the Lakekamu Basin. A complete inventory could not be accomplished in the time frame of the RAP survey. However, it is hoped that these data will serve as a beginning for continued study.

It appears that roughly half of the mammal species in New Guinea need help to ensure their continued existence; of these, 85% are endemic.

Methods

Trapping

Six trap-lines (four in alluvial and two in hill forest) each consisted of 40 trap-sites spaced 20 m apart on pre-cut trails and transects (See map page 22). For each line, we used 80 38 X 10 X 12 cm Sherman live-traps with open front doors and stationary wire-mesh back doors and 20 82 X 23 X 23 cm Tomahawk double-door live traps. Each trap-site had one Sherman on the ground and one Sherman placed 1 to 2 m above ground on a liana or tree. Alternating trap sites (every 40 m) had one Tomahawk trap placed on the ground. We used rolled oats and peanut butter as bait in the Sherman traps and used crackers with peanut butter, dried fruit, and fish as bait in the Tomahawk traps. Each line was run for 5-6 nights for a total of 2480 Sherman and 620 Tomahawk trap-nights. For each captured mammal, we recorded species, date, sex, age, head-body length, tail length, hindfoot length (with and without claw), ear length, reproductive condition, and mammary formula. We clipped dorsal fur and used colored beads sewn to the ears to detect recaptures. We collected 1-2 voucher specimens of each species.

Mist-Netting

We ran 39 mist-nets from 1-3 nights on 4 net-lanes, all in alluvial forest (See map page 22). We ran nets for a total of 7 nights and 12,965 net-meter-hours. For each captured bat, we recorded species, date, time, sex, age, head-body length, tail length, hindfoot length, ear and tragus length, forearm length, tibia length, and reproductive condition. We used forearm bands to detect recaptures and collected 1-2 voucher specimens of each species.

Harp Traps

We used two free-standing 2 X 2 m two-layer harp traps for 32 nights on Bulldog trail and for 12 nights on the helipad trail for a total of 88 harp-trap-nights. Bat measurements, recaptures and specimens were taken as above.

Identification and Specimens

All species were identified using Flannery (1995) and Dr. Frank Bonaccorso's unpublished keys for NG rodents and NG bats (PNG National Museum, P.O. Box 5560, Boroko, NCD, PNG). Voucher specimens are deposited at the PNG National Museum in Port Moresby and at the Academy of Natural Sciences in Philadelphia, PA, USA. For each specimen collected, we also prepared DNA tissue samples in buffers that will be deposited at the Academy of Natural Sciences in Philadelphia, PA, USA.

Results

Capture Rates

We trapped a total of 121 individual non-volant mammals with 80 recaptures; there were 6.5 captures (3.9 individuals) for every 100 trap-nights. We netted 52 individual bats with 3 recaptures; for every 1000 net-meter-hours we captured 4.2 bats. In the alluvial forest we, caught 14 bats in the harp traps with no recaptures (2.2 bats per 10 harp-trap-nights). We caught no bats in the harp traps in hill forest.

Density

We trapped four times as many *Rattus leucopus* as any other non-volant mammal (60 individuals). The next most trapped species were *Melomys levipes* (14), *Stenomys verecundus* (12), *Melomys rufescens* (11) and *Melomys platyops* (10). We found low numbers of *Uromys caudimaculatus* (5), *Melomys leucogaster* (4), *Lorentzimys nouhuysi* (3), and *Hydromys chrysogaster* (2). We found only one *Antechinus melanurus*, one *Pogonomys loriae* and one *Pogonomys macrourus*.

Syconycteris australis was captured four times more frequently than any other bat species (33 individuals). Next plentiful were *Nyctimene albiventer* (9), *Aselliscus tricuspidatus* (8) and *Paranyctimene raptor* (6). We found low numbers of *Hipposideros cervinus* (4), *Hipposideros muscinus* (3) and *Rhinolophus euryotis* (3). We found only one *Hipposideros diadema* and one *Pipistrellus wattsi*.

Diversity

We captured one marsupial, twelve rodent species, and eight bat species. We saw but did not capture two additional marsupial species and one additional bat species for a total of 24 mammal species (Appendix 17).

Alluvial and hill forest trapping both reached a species asymptote at about one week of trapping with 8-9 species captured (Appendix 1H) (additional trapping did not reveal new species). Both forest types shared six trapped species (four *Melomys*, one *Rattus*, and one *Stenomys*), but in alluvial forest we also trapped *Uromys caudimaculatus*, *Lorentzimys nouhuysi* and *Antechinus melanurus* and in hill forest we trapped *Hydromys chrysogaster* and *Pogonomys macrourus*. One additional rodent species, *Pogonomys loriae*, was captured from its underground burrow uncovered while we were digging a ditch.

The mist netting species/effortcurve asymptotes at 7000 net-meter-hours with six bat species (Appendix 1I) (further mist netting did not reveal more species). All mist netting was done in alluvial forest. We captured six bat species in the harp traps; four were also captured in the mistnets (*Syconycteris australis*, *Paranyctimene raptor*, *Rhinolophus euryotis*, and *Hipposideros muscinus*), two were only caught in the harp traps (*Hipposideros cervinus* and *Aselliscus tricuspidatus*). The harp trap species/effort curve reached an asymptote at nine days with five species, but one more species was captured 16 days after that. All harp trap captures were in alluvial forest; the 12 days in hill forest produced no captures. *Hipposideros cervinus* was also captured by batting it out of the air. One additional bat species, *Pipistrellus wattsi*, was captured from its roosting position in a tree 7 m above ground.

Reproduction

Adult females of *Hipposideros cervinus* (3), *Hipposideros muscinus* (2), *Hydromys chrysogaster* (1), *Stenomys verecundus* (1), *Nyctimene albiventer* (5), and *Paranyctimene raptor* (1) showed no sign of reproductive activity (sample size in parentheses). Eight rodent and four bat species were reproductively active during the study period (see Appendix 18).

Discussion

Density

Rattus leucopus and *Syconycteris australis* were the most common species encountered. Relative abundances of *Uromys* and *Hydromys* were probably higher compared to smaller species than suggested because they were too large for 80% of the traps. Furthermore, the relative densities recorded in this paper should not be taken as the relative densities of these species at Ivimka, but rather as the relative ease of capturing these species using our methods during the months of the study. These figures could be used to indicate the relative ease of conducting research with these animals at Ivimka.

Sampling Effort to Find Diversity

A minimum of 14 nights of trapping (7 in alluvial and 7 in hill forest at 100 traps/night), 9 nights of harp traps (2 traps/night), and 5 nights of mist-netting (7000 net-meter-hours) would have yielded all but one of the species we found over the 31 trapping nights (100/night), 44 harp trap nights (2/night), and 7 netting nights (13,000 n-m-h). So to find the species diversity we did with the methods we used, a team would need to spend at least 11 days for a single habitat or 3 weeks for two habitat types at one location (the extra days are needed to set, move, and wash traps). Surveys for mammals in New Guinea should be at least this long and ideally should be repeated during different seasons.

Alluvial Versus Hill Forest

Although the elevational difference in the alluvial and hill forests sampled was only 100 meters, three mammals species were captured in alluvial forest that were not captured in hill forest and two other species were captured in hill forest that were not captured in alluvial forest. We suspect that these differences were due to chance; two of the species (*Antechinus* and *Pogonomys*) were only captured once and the two water-rats (*Hydromys chrysogaster*) captured were in hill forest and not near the large river in alluvial forest as one would expect. Furthermore, *Lorentzimys nouhuysi* and *Uromys caudimaculatus* are typically found up to 2700 m and 2000 m elevation respectively; there is no reason for them to be in 120 m alluvial forest and not in 250 m hill forest.

Reproduction
Five of the adult female *Melomys levipes* had the normal mammary formula of 0+2=4, but one individual had a formula of 0+3=6. This variation has not been recorded in the literature. This appears to be the first record of reproduction for this species (Flannery 1995).

Additionally, one adult-sized female *Melomys levipes* was definitely non-perforate, yet was heavily lactating. This suggests that either the hymen is capable of growing back, or that perhaps some young adult females aid in lactating the progeny of other females before they conceive litters of their own. Virgin goats can be induced to lactate by treatment with estrogen and progesterone or by simply stimulating the nipples with the mechanical action of suckling (Cowie 1984). Lactation in male bats may be brought about by transdermal transference of estrogens through communal roosting (T. Kunz, personal communication). Communal nesting by rodents could stimulate lactation in sub-adult individuals (either from nestlings suckling or from transdermal transference of estrogens from lactating adult females) who could then help with lactation chores. Looking into this further may be a fruitful research project at Ivimka.

Melomys levipes and *Melomys rufescens* both had independent juveniles, lactating adult females, pregnant adult females, and non-reproductive adult females captured; this suggests aseasonal reproduction in these species.

Melomys platyops was previously known to lactate in January (Flannery 1995) and we found individuals lactating in November, thus extending the known reproductive season.

Lorentzimys nouhuysi was pregnant in October; it has previously been recorded pregnant in December and in July (Flannery 1995).

Flannery (1995) recorded *Pogonomys macrourus* with dependent juveniles in July and in October; the individual we captured was lactating in November.

All other species we found lactating or preg-

nant have previously been recorded as being reproductively active during the time period we found them, and embryos counts for all pregnant females were as previously established in the literature (Flannery 1995).

Other Interesting Points
We had two color morphs of *Aselliscus tricuspidatus*; some individuals had bright orange fur and others had dull brown fur; sex was not a predictive factor in this polymorphism. Ammonia fumes from concentrated guano in roosting caves can bleach a bat's fur orange (Fleming 1988). The orange bats are possibly roosting in a cave with concentrated ammonia fumes and the brown bats are not.

Our captures of *Melomys levipes* represent a 150 km northwest range extension, the nearest previous records being from the Sogeri Plateau/Astrolabe Range (Flannery 1995).

Of the 24 mammal species verified to occur at Ivimka, one is data deficient (*Melomys levipes*), two are vulnerable (*Paranyctimene raptor* and *Pipistrellus wattsi*) and one is threatened (*Hipposideros muscinus*) (Appendix 17).

Conservation Recommendations
The most obvious recommendation is to limit, or regulate hunting of mammals. This will be difficult for local people to sustain as, in many cases, game is a primary source of protein. Perhaps local hunters would agree to a ban on a large area around the research station if they can still hunt in other areas. This would give a protected area for game populations (source) that could replenish adjacent regions that are hunted (sinks). A ban on hunting around the station would enhance the attractiveness of the study area to visiting scientists and enable research into the ecology and demography of mammal species. Furthermore, with a control area where hunting is prohibited, comparisons could be made in future years with hunted areas to assess the impact of hunting and evaluate the effectiveness of the conservation initiative.

The second recommendation is also obvious. More research on various species— especially those that we know so little about and that are likely to have low reproductive rates or high mortality rates. Unless we know more about these animals, we cannot intelligently protect them.

A study on mammal use by local people should be initiated. Learn which species are taken, how many individuals of each species are taken, what times of year they are taken, the age and sex of individuals taken, and the methods used to capture them. Find out if hunting patterns are changing over time with the introduction of new technology or with the changing preferences of local people (have ongoing wildlife-use studies). We must learn use patterns to know if they are sustainable or if they need to be altered in some cases.

SOCIAL HITORY OF THE LAKEKAMU RIVER BASIN (Stuart Kirsch)

Chapter Summary

• There are four major landowner groups claiming ownership to lands and utilizing resources within the Basin: the Biaru, the Kamea, the Kovio and the Kurija.

• The Kovio are the only long-term residents of the lowland part of the Basin. They traditionally controlled the shell trade from the coast into the highlands.

• Shifting agriculture, as practiced in the Basin, has had little impact on the overall vegetation because it is restricted to fertile floodplains near the major rivers.

• Sale of Betelnut is primary source of revenue, followed by small-scale gold extraction, and wild game.

• Gold mining shows promise in some areas of the Basin, but large-scale dredging operations have been precluded due to land ownership disputes.

• Traditional boundaries in the Basin are unclear and contentious.

• There is some interest in attracting major, large-scale resource development projects to the Basin, particularly among some of the Kovio; this could pose an obstacle to the conservation initiative.

Introduction

Environmental conditions in the Lakekamu Basin have been affected by centuries of human occupation and use. Research on the social history of the Basin is thus a necessary complement to the analysis of local biological diversity. This information is also integral to proposed conservation initiatives in the region (Kirsch 1997). There is an urgent need to connect analyses which "emphasize traditional and internal factors as an explanation for human activity" with studies that address "external constraints of an evolving global or regional political economy." (Filer 1994: 189). Elsewhere, I have described the major prospects for economic development under consideration by the inhabitants of the Lakekamu Basin (Kirsch 1997). In this chapter, I focus on historical patterns of land and resource use.

The resources of the Lakekamu Basin have long been exploited by the four cultural-linguistic groups living on its margins (Appendices 19 and 20). The Biaru live in the northeastern foothills of the region and are related to other Biaru-speaking groups in Morobe Province to the north. The Kurija speak the language of the Kunimaipa, who live in the mountains of Central Province to the east. Prior to settling in the eastern edge of the Basin, the Kurija lived near the headwaters of the Kunimaipa River. The Kamea live in the mountains in the northwest corner of the Basin. They belong to the Angan-speaking populations who dominate the eastern portion of PNG's central mountain range (Bamford 1997). Of the four groups, only the Kovio have a long history of residence in the lowlands. They moved from the southern part of the Basin to its western edge in the 1950s though the exact dates are disputed due to their implications for land claims (Filer and Iamo 1989). In describing the region, Filer and Iamo (1989: 17) point out that, "Even by Melanesian standards, the. . . area exhibits a measure of cultural diversity which is truly remarkable, and the demarcation disputes which characterize the relationships between cultural groups have a long and complex history."

Research on the social history of the Basin is thus a necessary complement to the analysis of local biological diversity.

Contemporary Economic Conditions

Most of the two thousand residents of the Lakekamu Basin currently live in villages close to the government center at Kakoro, in the northern half of the Basin. Several other villages are located along the major river courses and in the mountains to the northwest. The name Kakoro means "dried up" or "hungry" in Motu, the trade language used throughout Papua. Founded in 1972, Kakoro has a school, a health center, an office for a provincial administrator, a small market and a grass airstrip. The southern half of the Basin remains largely uninhabited, although it is regularly visited for hunting, fishing and for cutting timber and gathering other forest products.

The Biaru, Kurija and the Kamea plant large, swidden gardens, usually along river banks, in which they grow bananas, sweet potato, taro "tru" and taro "kongkong", "aibica" and other greens, sugarcane, pitpit and yams. June and July are the months of the yam harvest and the dry season runs from September to December, which are the best months for hunting. The three groups make limited use of sago (*Metroxylon* sp.), which is a staple food throughout much of the interior lowlands of New Guinea (Ruddle *et al.* 1972). Their most important tree crops are betelnut, okari, breadfruit and *marita* pandanus. The Kamea trade lowland produce for *karuka* (nut) pandanus, which is grown at higher altitudes. The Kovio have a more typical lowlands subsistence economy than their neighbors, which is based on hunting and fishing, harvesting sago and planting modest-sized gardens.

Residents of the Lakekamu Basin can sometimes sell garden and forest products to government personnel in the Basin. Peanuts and betelnut are also sold in urban markets in Kerema, Wau and Port Moresby, where a large bag of betelnut is worth up to K200. Harvests of coffee and cocoa, planted throughout much of the area, have been limited because of transportation problems (Filer and Iamo 1989). While avocados grow well in the Basin, they have not been exploited commercially for the same reason. Okari nuts (in season), meat from wild pigs, crocodile skins and feathers (from birds of paradise, sulphur-crested cockatoos, *Goura* pigeons, and cassowaries) are also sold in the markets of Port Moresby. Much of the money earned on trips to urban markets is spent on travel and other expenses, although people usually return home with some durable goods, such as kerosene lanterns, cooking pots or clothing. A few residents of the Basin earn small sums of money by selling colorful beetles and butterfly larvae to the insect-marketing program sponsored by the Wau Ecology Institute (Orsak 1991). It would be useful to monitor or sample how much trade in forest products actually transpires.

Transportation in and out of the Basin is expensive and unreliable. The plane to Wau cost K55 one way in 1996. Although scheduled to arrive twice a week, these flights are often postponed. It is also possible to reach Port Moresby by travelling south along the Lakekamu River by motorcanoe to its junction with the Tauri River, where a road links the settlement of Iokea to the city. In 1994, this canoe trip cost K30, with a surcharge of K5 per large bag of betelnut. Passage along the road cost an additional K10.

Some of the Basin residents are involved in small-scale alluvial gold mining in the creeks that run through the hills north of Kakoro (Filer and Iamo 1989). The Biaru have panned for gold in Nowi creek for more than a decade. The residents of a settlement known as Bundi Camp, located just north of Kakoro on the Biaru River, migrated into the Basin from Goroka (Eastern Highlands Province) to work local gold deposits. The Kurija residents of Mirimas village pan for gold in the eastern tributaries of the Oreba River (Filer and Iamo 1989). Some of the residents of the Kamea village of Iruki migrated to the Basin to work the gold at Omoi Creek (also known as Cassowary Creek). In the late 1980s, the Biaru invested in dredging equipment, but this angered the Kurija, who claim ownership of the land beside Nowi Creek. With support from the Kovio, the Kurija raided the Biaru mining camp, destroying their equipment and causing a number of injuries. The Biaru were backed up in the dispute by the coastal Moveave, who have their own territorial quarrels with the Kovio. Both sides went to court in 1993 and a temporary injunction was placed on mining until the land disputes are settled, although small-scale gold panning continues unabated.

CONSERVATION INTERNATIONAL

More substantial gold reserves have been located near the Olipai River, a western tributary of the Lakekamu River. Filer and Iamo carried out a base-line planning study for the Lakekamu Gold Project in 1989. They describe a mining project with a large dredge capable of processing several million cubic meters of material a year during the projected five-fifteen year life-span of the ore body. If implemented as planned, Filer and Iamo predict that the project will have a "very *limited impact on the natural environment*" (1989:15; emphasis in the original). The project would also create employment opportunities for residents of the Basin and improve access to urban markets. To date, the project remains on hold. Several mining companies continue to prospect for additional ore deposits in the mountains north of the Lakekamu Basin (C. Makamet and J. Sengo, personal communication).

Timber is another potential resource. The government has not granted any logging concessions in the Lakekamu Basin, although a proposed timber project in Gulf Province would encompass the southwest corner of the Basin (T. Werner, personal communication). Several Malaysian timber companies have expressed interest in obtaining logging rights for the remainder of the Basin (B. Beehler, personal communication). Logging might also be introduced into the area indirectly, by establishing an oil palm plantation organized into a series of small-holdings, although the residents of the Basin have already rejected one such proposal (C. Makamet and J. Sengo, personal communication). In 1994, the residents of the Lakekamu Basin were divided about the prospect of logging projects on their land.

Fieldwork and the Politics of Culture

This study is based on several weeks of ethnographic research in the Lakekamu Basin. I visited most of the villages in the vicinity of Kakoro and held meetings and discussions (in Pidgin English, assisted by C. Makamet) with members from each of the four groups living in the Basin. The following discussion of disputed borders and territories is meant to illustrate important causes of local conflict and tension, rather than to resolve

them; more complete documentation of the history of the Lakekamu Basin will no doubt reveal the shortcomings of this preliminary analysis. It is unlikely that a single "true" history of the disputed lands will ever emerge and settle the existing problems associated with the dispute (Filer and Iamo 1989). This report should not be used in court proceedings regarding land rights in the Basin.

Land rights in the Lakekamu Basin are subject to heated and sometimes violent local debate, affecting how people talk about themselves and the past. Political and economic changes during the last few decades have put considerable pressure on "traditional" ways of organizing social relations and access to resources. Even seemingly basic concepts such as "landownership" are the product of complex interactions between local systems of land tenure and resource rights on the one hand, and the legal categories and capitalist economy of the state on the other (Filer 1997). What anthropologists call the "politics of culture" (Linnekin 1992), referring to how political and economic forces shape culture, identity, and claims made about the past, is of paramount importance in the Lakekamu Basin.

The Biaru

The Biaru live in several villages north of Kakoro, on the west side of the Biaru River. These settlements are Kakoro village (adjacent to the government center), Amamas Camp, Poian Camp and Meri Camp. There are four major Biaru descent groups or patri-clans (see Appendix 21), although clan membership is no longer of primary importance and the Biaru say that such affiliations have become "all mixed up."

The Biaru own the land in the mountains northeast of Kakoro. They also claim the land from the east side of the Biaru River to the Avi Avi River in the west, which they describe as the border separating their land from that of the Kamea. In contrast, the Kamea maintain that they own all of the land between the two rivers. The difference in perspective is related to the history of the Basin. Before the imposition of colonial control, the Kamea used to raid Biaru settlements

Land rights in the Lakekamu Basin are subject to heated and sometimes violent local debate...

(Filer and Iamo 1989), effectively preventing the Biaru from settling in the disputed territory. The Biaru were intimidated by the Kamea, whom they accused of practicing cannibalism (the attribution of cannibalism to one's neighbors is common throughout New Guinea, and is often a metaphor for social relations rather than a literal statement about dietary practices). Both groups probably made intermittent use of the resources of the area between the Biaru and the Avi Avi Rivers, although neither group settled there.

The Biaru also maintain that the land along the Si and Nagore Rivers, south of Kakoro, belongs to them, although both the Kurija and the Kovio dispute this assertion. The Biaru argue that the Kurija have no land rights in the area, claiming that they only recently moved into the Basin from the headwaters of the Kunimaipa River, a distance of several days walk. At issue is the timing of the Kurija migration into the lowlands. In contrast, the Biaru acknowledge the presence of the Kovio, with whom they had long-standing trade relations.

Biaru leaders were reluctant to speak with me about the history of the area because of the ongoing dispute over the gold at Nowi creek. Mindful of the unsettled issues surrounding land ownership in the Basin, they told me the following myth, which describes their relationship to the other Biaru (or Gorua) speakers in the mountains of Morobe Province:

Sankep is the name of the ancestor of the Biaru [who live northeast of Kakoro]. He came to the Biaru River with his son Tie. Sankep did not know how to build a house, so he slept in the forest. Tie put his first born son Paur in a men's house and decorated him with the shells that they used in exchange. Men came to the ceremony from all around.

Moin came from the headwaters of the Si River. He killed a pig, took off its skin and came down from Maoru mountain. Along the way he met a man called Maorarai, and the two camped there overnight. Maorarai wanted to kill Moin, but someone warned Moin and he ran away. Moin then sent a message: when the river becomes dirty, Tie should climb to the mountain top. When the time came, women in the mountains pissed into the river, fouling the water downstream. This told Tie that it was time for him to climb the mountain.

Tie and his wife Ruipispis started walking to Maoru mountain. Tie stopped to look for game in the forest, and sent his wife ahead along the track. She met with Moin

along the way and they had sex together near a small patch of bamboo.

When Moin and Tie met, Moin showed Tie how to build the men's house, which they call aniak. Moin showed Tie how to make a knife from bamboo. He also taught Tie about exchange, and as a result, they became friends and trading partners.

The myth addresses history, social relations and land rights and has a broad regional distribution; Moveave versions of the myth explain inter-group relations on a wider scale (Filer and Iamo 1989). It describes the relations between the two Biaru groups: the main population in the mountains of Morobe Province, and the splinter group that moved into the foothills to the south, close to Kakoro. According to the narrator of the myth, the two groups share one language, but maintain separate descent groups.

Sankep and his son Tie represent the Biaru living on the edge of the Lakekamu Basin. The dispute between the two groups, represented by Moin's affair with Ruipispis, Tie's wife, is resolved through exchange. Moin gives Tie access to sacred knowledge, enculturating him through participation in male cult ritual. The two groups are also linked into an exchange relationship through their joint participation in the ritual. The events depicted in the myth follow a common Melanesian scenario in which exchange transforms conflict into productive relationships (Schieffelin 1976).

The myth also helps to explain the dynamics of regional exchange in the Lakekamu Basin. The lowland Biaru occupy an important position in this system. In the past, they had regular trade relations with the Kovio, who controlled exchange between the coast and the highlands (see Hau'ofa 1981 on the neighboring Mekeo). This was an important route for shell valuables central to the exchange economies of the highlands. The shells were obtained from the coast and from lowland mangrove forests. They were traded from the south through the Lakekamu Basin, becoming increasingly valuable with their distance from their source. In return for shells, the Biaru traded bird feathers, bark cloth and spears to the Kovio. From the mountain Biaru, they acquired pigs and quarried stone used for axe

CONSERVATION INTERNATIONAL

Rapid Assessment Program

blades. Shells were central to Biaru exchange; they explained that they used them "like money" and in bridewealth payments.

The Kurija

The Kurija are the western-most group of the Kunimaipa, who live in the mountains of Goilala sub-district in Central Province (Hallpike 1977). Their language has three major dialects: the first is spoken in the mountains, the second near the headwaters of the Kunimaipa River and the third among the Kurija. In the 1950s, Australian anthropologist Margaret McArthur worked in the upper Kunimaipa valley. What the Kurija told me about their social organization and ritual practices was consistent with what McArthur (1971) described for the highland Kunimaipa.

The Kurija are patrilineal, with their land held in trust by the lineage, rather than divided among its members. People from outside the lineage are not permitted to use these resources unless permission has been explicitly obtained. This includes use rights for gardening, fishing in rivers and streams, and hunting in their forests. If a person plants a tree on another person's land, the landowner may uproot it.

Kurija is known as the "big name" for their clan structure, which they describe as the "backbone" of their society. There are nine Kurija lineages, two of which are now defunct (see Appendix 22). The lineages are named after their founder and are exclusively local. Each has a headman, known as *amip*, who directs its affairs. The position is acquired through hereditary primogeniture. Headmen are expected to spend much of their time socializing and telling stories. During feasts they stand up and instruct participants to refrain from stealing, fighting or otherwise inciting conflict. The main role of the headman is to settle disputes that arise between lineages or opposing clans, leading the Kurija to compare his duties to those of a lawyer.

The Kurija explained that their lineages did not fight among themselves in the past because their numbers were so small. Disputes were settled by the payment of compensation, such as the exchange of a pig, which was generally negotiat-ed by the respective headmen of the two lineages. In the case of a dispute with members of another group, however, all of the Kurija would join together to fight if necessary.

Kunimaipa lineages are exogamous and after marriage the couple resides patrilocally. Both sides of the family are expected to make small gifts of dogs' teeth and bird of paradise feathers, but until recently, no large exchange of wealth was associated with marriage. Today, bridewealth payments of K500-K1200 are usually expected (Appendix 23).

The following myth explains how the Kurija settled in the Lakekamu Basin. The Kurija used to live in the mountains east of Kakoro, at the headwaters of the Kunimaipa River, which they call the Kutkut. There they belonged to the clan called Komi Garoi, named after the senior man of the group. A dispute between two factions of the clan caused a permanent rift to emerge:

The people known as the Komi Garoi took turns hunting and collecting firewood. When the people responsible for bringing firewood back to the village forgot to do so, it led to a quarrel. The dispute was settled by splitting into two groups, the Komi Garoi and the Kurija.

The Kurija left the mountains and moved to the head-waters of the Nagore River. In all, seven brothers settled along the Nagore. Only men moved down from the mountains; there were no women.

One evening, one of the men saw something odd and called the others to come and look: there was a strange light coming from a stand of bamboo. From a distance, it looked like fire. "What is it?" they wondered. They decided to go closer and investigate. They also sent a message to Komi Garoi in the mountains, inviting him to come.

When they reached the bamboo, they saw that the light inside was made by women. One by one the women stepped outside, and the men began to choose the women they preferred: "you take this one, I'll take that one," and so on. Each man chose one woman as they came forward.

Komi Garoi came down from the headwaters of the Kutkut River. He told the men, "Now you are married, so it is time to have a marriage feast." Before they ate, how-ever, Komi Garoi stood and spoke again: "Now I'll divide the land among you." He designated a different plot of land for each brother. The brother named Lairao (of the narrator's clan) was to live on the land between the Si and Nagore Rivers. The eldest brother, Kurija, was to live along the Kunimaipa River. The other brothers were told to live in different locations along the Si, Nagore and Kunimaipa Rivers.

Now Komi Garoi is dead; his body is on the other side of the Kunimaipa, at a place called Kuiki (Sharp Mountain).

The myth operates at several levels. It asserts that the Kurija have lived in the lowlands between the Kunimaipa, Si and Nagore Rivers for many years. One genealogy that I collected suggests that six generations of the Kurija (about 150 years) have lived in the Lakekamu Basin, although this is not an accurate means of making chronological assessments, for such genealogies are easily telescoped or foreshortened. The myth also explains the affiliation of the Kurija lineages and the distribution of land between them. There is more to the myth than a group of people staking out land claims, however; it also explains the relations between the highland and lowland Kunimaipa.

Conflict between clan members may be resolved through fission, separating the feuding parties into independent groups. The event in the myth that precipitated the split is trivial in comparison to its consequences, however, suggesting that another interpretation is necessary. The alternation of tasks between the two lineages suggests a society with a simple division of labor. Physical labor was reciprocated directly, in contrast to more complicated (and more prevalent) systems in which the products of labor are exchanged directly through barter, or indirectly by means of mediating items of value, such as shells or money. Yet the failure of one group to collect firewood as expected reveals the insufficiency of this mode of exchange.

Komi Garoi is credited with instructing the Kurija to move from the mountains into the lowlands. This allowed the Kunimaipa to establish trade relations with the Kovio for shells from the coast. In return for bows and arrows, tobacco and dogs' teeth, the Kurija acquired shell valuables which they later traded to the Kunimaipa living in the highlands. This enabled the Kunimaipa to establish a more complex division of labor based on a system of inter-group alliances, which were organized through the exchange of shell valuables in bridewealth and other transactions.

The second half of the myth is concerned with the relations between the Kurija and the Kunimaipa who remained in the mountains. When the women emerge from the stand of bamboo, the resulting marriages are anomalous. Since the women have no relatives, the Kurija have no

affines with whom they can form alliances. Komi Garoi approves the marriages nonetheless and apportions the land accordingly. The Kurija are no longer beholden to their relatives in the mountains for either land or marriage; the myth asserts their autonomy. The Kurija did not resume intermarriage with other Kunimaipa speakers from the mountains until recently. The myth describes the movement of the Kurija from the mountains into the Lakekamu Basin, where they became an independent group that participated in the shell trade from the coast into the highlands.

The Kurija claim the land between the Biaru and the Kunimaipa Rivers, including the territory between the Si and Nagore Rivers. They confirm the Biaru assertion that the two groups did not have contact until recent years, although they claim that the Biaru River is the border between the two groups. Like the Biaru, the Kurija deny having had contact with the Kamea before the colonial period, or having ventured into their territory. They only saw footprints that the Kamea left behind. They traded regularly with the Kovio, sometimes learning to speak their language, although the two groups did not intermarry. The Kurija do not understand the languages of either the Kamea or the Biaru.

The Kurija currently live in two villages, Totai and Mirimas, along Biaru River just south of Kakoro. They make their gardens in the floodplain of the river and along several of its smaller tributaries. A survey of recent marriages at Mirimas village suggests a relatively high rate of village endogamy.

In an interview about the future of the Lakekamu Basin, several Kurija leaders expressed their concern about large-scale development or resource extraction projects. They are opposed to large mining projects because of the chemicals (*marasin* in pidgin) that are released into local waterways, which can poison fish and other riverine life. They were concerned that the noise from generators and engines would scare off the wildlife and ruin their hunting. They also feared that the people brought from the outside to work on these projects might cause trouble and disrupt their way of life (*bagarapim ples* in pidgin). They opposed having large corporations

CONSERVATION INTERNATIONAL

operate in the Basin, although they were supportive of smaller-scale development projects that would not harm the natural environment (*spoilim ples* in pidgin).

The Kamea

The Kamea living on the edge of the Lakekamu Basin have close ties to the communities in the Kaintiba and Kamena regions of Gulf Province. Tekadu and Nukeva are the Kamea villages located closest to the Basin and the Ivimka Research Station. Marriage exchange in this patrilineal society took the form of gifts of smoked game and garden products. After killing and smoking a large quantity of game, a man interested in marriage would make a formal gift of meat to his prospective in-laws. If the offering was accepted, he would plant a new garden, build a new house and invite his affines to attend a feast. There were no bridewealth transactions involving shell valuables, although the Kamea living in the area near Kaintiba used shell valuables called nuwa in their bridewealth exchanges (S. Bamford, personal communication).

Given that this group of Kamea did not exchange shells, they had no need to participate in the regional exchange system to their south. Unlike the Biaru and the Kurija, they had no interest in the shell trade monopolized by the Kovio. Their exchange relations were oriented toward other Kamea communities living at higher altitudes, to whom they traded lowland produce. As a consequence, the Kamea had no pre-contact social relations with their neighbors in the Lakekamu Basin.

Kamea raids against the Biaru to their east prevented settlement in the northern portion of the Basin, from the Avi Avi River to the Biaru River (Filer and Iamo 1989:31). While both the Kamea and the Biaru made intermittent use of resources from this area, it remained unoccupied until the colonial era. Current disputes over land ownership reflect this history.

The Kamea village of Iruki was established by the residents of Nukeva and Tekadu after Kakoro was founded in 1972 (see Appendix 24). After primary schools were built in the mountain villages, most of the Kamea living at Iruki returned home. To maintain their settlement at Iruki, and to protect their rights to nearby land and resources, the Kamea invited people from outside of the Basin to settle in Iruki in their place. Most of the inhabitants of Iruki are from Kamena and Kaintiba in the mountains. Few of the people living in the village have local land rights.

Unlike other communities in the region, there is little village endogamy in Iruki. This is not surprising given its relatively transient population. Only a quarter of the people in a sample of recent marriages among village residents consider either Iruki or the two neighboring Kamea villages of Tekadu and Nukeva to be their home. At least half of the marriages today are accompanied by payments of cash (see Appendix 25).

The people living in Iruki village, although linguistically and culturally Kamea, do not own land in the Lakekamu Basin. They are "placeholders" for the residents of Nukeva and Tekadu. Although they have permission to exploit the resources of the Basin themselves, they do not have the right to decide its fate with respect to development or conservation projects.

The Kovio

The Kovio live in three villages on the western edge of the Lakekamu Basin, and in urban areas of Gulf and Central Province, including Kerema and Port Moresby. They speak an Austronesian language that is most closely affiliated with Mekeo (Brown 1973). The neighboring Mekeo are divided into two groups, the Bush Mekeo who live in the swampy areas south of the Lakekamu Basin (Mosko 1985), and the Central Mekeo who live to the southeast (Hau'ofa 1981; Stephen 1995). The Kovio maintain that they have strong cultural continuities with the Mekeo as well, including the ritual body and face-painting style that has made the Mekeo famous throughout PNG. The Kovio commonly intermarry with the Bush Mekeo, which gives them valuable access to resources outside of the Basin. These marriages also give the Kovio access to a wider range of "wantok" relations (literally "one talk," but referring to common origins and interests), which is particularly important for a group as small as the Kovio, with a population of less than five hundred.

The Kovio lived in the flat, swampy lowlands along the southern tributaries of the Kunimaipa River until they moved north to the junction of the Kunimaipa and Biaru Rivers, probably in the 1950s. When this settlement was flooded, the Biaru moved further west to the land adjoining the Kunimaipa and Tiveri Rivers, the current location of Okavai village. Urulau village is west of Okavai, near the border between Central and Gulf Provinces, and Ungima is a new settlement closer to the Mekeo. Unlike their neighbors, they have no village in the immediate vicinity of Kakoro, although several Kovio civil servants are employed at the government station.

At the time of the 1990 census, Okavai had 26 households and 175 residents (PNG National Statistical Office 1993). The village itself is divided into five exogamous patrilineages (see Appendix 26). In the past, each lineage had a pair of authority figures: a political leader or headman, and a sorcerer (Stephen 1995). Today, only the headman of each lineage is recognized. Two of the five lineages, Unga and Kongopu, are particularly prosperous, and many of their members live and work in urban areas. The two lineages also have the most political influence.

The Kovio claim ownership of most of the land in the Lakekamu Basin. Some of the Kovio suggest that their land rights extend north along the Tiveri River, past its junction with the Avi Avi River, and into the mountains of Morobe Province, near the colonial border between Papua and New Guinea. This includes all of the land in the Basin claimed by the Kamea. Other Kovio deny that their land rights extend that far north. Given the friction between the Kovio and the Kamea in pre-colonial times, it is not surprising that the boundary between them remains in dispute. The Kovio also claim the land along the Si and Nagore Rivers. They challenge Kurija claims to this area, arguing that Kurija land rights are limited to the upper Kunimaipa River. According to one account, a Kovio man working with an expatriate pastor from the United Church invited the Kurija to settle in the Basin, but did not trans-

fer ownership of the land to them.

Writing about the neighboring Mekeo, Hau'ofa (1981:17) notes that they were well-positioned to manage the trade throughout the region:

With the Kuni and Goilala, the Mekeo traded not only betelnut but also the shell ornaments they had received from the coast, in return for stone axes, flints, plumes and Pandanus nuts. Being located strategically along a major waterway between the coast and the mountains, Mekeo acted as intermediaries through whom the products of the sea and the mountains passed both ways.

The Kovio were similarly positioned with respect to their neighbors in the Basin. The Kovio traded shells along with bows and arrows to the Kurija in return for dogs' teeth (a local valuable as well as a ceremonial display item) and pigs. They also had established trade relations with the Biaru, who had a similar appetite for the shells they controlled. In their analysis of social relations in the middle Sepik, Gewertz and Errington (1991) suggest one group may dominate even in a system of "commensurate differences" in which all societies are interdependent and relatively equal, one society may still dominate the others. The Kovio ability to control the shell trade in the Lakekamu Basin is reflected in their sweeping land claims. Their contemporary political influence recapitulates these patterns.

Their relations with the Mekeo and the location of their villages along the Lakekamu River give the Kovio greater access to urban resources. They are probably the best educated of the four groups in the Basin. They are actively involved in regional politics. They are also more "fully integrated into the cash economy" than their neighbors (Filer and Iamo 1989:54). A landowner association established by a lawyer from Okavai village is said to represent all of the Kovio, as well as other people living in the Basin. The association's interest in attracting large-scale resource development projects to the Lakekamu Basin may be the greatest obstacle to the successful implementation of the Lakekamu Conservation Initiative (Kirsch 1997). The urban orientation of the Kovio leadership may make them more willing to develop the natural resources of the Lakekamu Basin than the other residents of the Basin, who depend upon the local environment for subsistence.

Environmental History of the Lakekamu Basin

The lowland resources of the Lakekamu Basin have been regularly exploited by the four cultural-linguistic groups who previously lived on its margins: the Biaru, the Kurija, the Kamea and the Kovio. The first three of these groups are segments of highlands populations who moved into the Basin to take advantage of lowland resources and exchange networks. The fourth group, the Kovio, are the only long-term residents of the lowlands. Due to their strategic position in the south of the Basin and their close relations with the neighboring Mekeo, the Kovio were able to dominate the shell trade from the coast into the mountains. They achieved this by forming exchange partnerships with the Biaru and the Kurija, who were conduits for the shell trade to their mountain relatives. The Kamea remained independent of this regional exchange sphere and were widely feared by the other groups in the Basin, against whom they sometimes staged raids and fought.

Land rights in the Lakekamu Basin are complicated by competing and overlapping claims. Much of the northern half of the Basin was once an uninhabited buffer zone between the Kamea and the Biaru. The resources of this area were exploited by both groups, but their hostile relations made settlement in the lowlands too dangerous. When the government center at Kakoro was established in 1972, however, the northern half of the Basin became home to almost all of its inhabitants. This higher population density translated into greater environmental impact in the area surrounding these settlements. The shifting horticulture of the Basin dwellers has not affected as great an area as it might have, however, for their gardening is largely restricted to the fertile river floodplains, rather than spreading out through the adjacent forest land.

The pattern of human occupation in the southern half of the Basin is almost the reverse. It is not clear how long ago the Kurija migrated into the southeastern portion of the Basin, although estimates range from several generations to several hundred years. The Kurija have asked that archaeologists survey the area in order to prove the antiquity of their claims. The land between the southernmost hills of the Basin and the Kunimaipa River used to be occupied by the Kovio. Given their low population density and their emphasis on hunting, fishing and sago, they probably had very limited impact on the natural environment. Although they migrated from this area in the 1950s, they still return to hunt, often with shotguns, and usually for large game, including cassowaries, wild pigs and crocodiles. The southern half of the Basin is still dominated by primary forest, although the long-term effects of previous occupation by groups no longer resident in the Basin, such as the Moveave, could still be manifest (Filer and Iamo 1989). The establishment of Kakoro in 1972 left the entire southern half of the Basin completely uninhabited.

I have described the social history of the Basin in a relatively schematic fashion. For a more detailed discussion of population movements, see Filer and Iamo (1989). Another limitation of this report is that I have not done justice to the range of views and experiences represented within each of the four groups. It should not be assumed that individual perspectives are consistent within each group, nor even that group members have the ability to reach a consensus on important issues that they confront. In addition, the historical depth of my analysis is limited. The first step towards rectifying this problem would be to review the historical records from the colonial era, a task complicated by arbitrary political boundaries that have separated the peoples of the Lakekamu Basin into a number of different provinces (Gulf, Central, Morobe and Madang) and territories (Papua, New Guinea and Papua New Guinea). A fuller determination of the time depth of settlement in the Basin would require archaeological survey work and excavation. Historical patterns of occupation may correlate with local biodiversity in significant ways, and natural scientists should take the social history of the Lakekamu Basin into consideration as they carry out their research.

Proposed conservation initiatives must also take the social history of the Basin into account (Kirsch 1997). Given the small scale and size of most political groups in PNG, lessons from this project can be applied throughout the country.

Conservation

projects must

help develop

new forums

for debate and

planning within

communities

and across

social

boundaries.

Conservation projects must help develop new forums for debate and planning within communities and across social boundaries Given the complexity of local social history, including raids, trade monopolies, migrations and land disputes, new strategies of organization are required to enable the residents of the Lakekamu Basin to work together to shape the environmental future of the region. Finally, social research carried out in the Basin must consider the impact of economic and political forces on local understandings of culture and identity, the significance of past events, and aspirations for the future.

LITERATURE CITED

Airy Shaw, H. K. 1980. The Euphorbiaceae of New Guinea. Hobbs the Printers, Southampton. 243 pp.

Allen, G. R. 1989. Freshwater Fishes of Australia. T. F. H. Publications, Inc., Neptune City. 240 pp.

Allen, G. R. 1991. Field Guide to the Freshwater Fishes of New Guinea. Christensen Research Institute, Madang. 268 pp.

Allen, G. R. 1997. *Lentipes watsoni*, a new species of freshwater goby (Gobiidae) from Papua New Guinea. Ichthyological Explorations of Freshwaters 8: 33-40.

Allen, G. R. and M. Boeseman. 1982. A collection of freshwater fishes from western New Guinea with descriptions of two new species. Records of the Western Australia Museum 10: 67-103.

Allen, G. R. and D. Coates. 1990. An ichthyological survey of the Sepik River system, Papua New Guinea. Records of the Western Australia Museum. (Supplement) 34: 31-116.

Allen, G. R., L. R. Parenti, and D. Coates. 1992. Fishes of the Ramu River, Papua New Guinea. Ichthyological Exploration of Freshwaters 3: 289-304.

Allison, A. 1993. Biodiversity and Conservation of the Fishes, Amphibians, and Reptiles of Papua New Guinea. In Beehler, B. M.(ed.). Papua New Guinea Conservation Needs Assessment, Vol. 2: 157-225. PNG Department of Environment and Conservation, Boroko.

Allison, A. 1996. Zoogeography of amphibians and reptiles of New Guinea and the Pacific region. In Keast, A. and S. E. Miller (eds.). The origin and evolution of Pacific Island biotas, New Guinea to eastern Polynesia. Pp. 407-436. SPB Academic Publishing, Amsterdam.

Archbold, R. and A. L. Rand. 1935. Results of the Archbold Expeditions. No. 7. Summary of the 1933-1934 Papuan Expedition. Bulletin of the American Museum of Natural History 68: 527-579.

Bakker, K. and C. G. G. J. van Steenis. 1957. Pittosporaceae. Flora Malesiana ser. 1, 5(3): 345-362.

Bamford, S. 1997. The Containment of Gender: Embodied Sociality among a South Angan People. Ph.D. dissertation, University of Virginia. 223 pp.

Barker, W. R. 1980. Taxonomic revisions in Theaceae in Papuasia. I. *Gordonia, Ternstroemia, Adinandra* and *Archboldiodendron*. Brunonia 3: 1-60.

Bauer, A. and K. Henle. 1994. Familia Gekkonidae (Reptilia, Sauria). Part 1 Australia and Oceania. Das Tierreich 109: 1-306.

Barber, S.A. 1995. Soil nutrient bioavailability: a mechanistic approach. John Wiley and Sons, New York. 414 pp.

Beckon, W. N. 1992. The giant Pacific geckos of the genus *Gehyra*: morphological variation, distribution, and biogeography. Copeia 1992: 443-460.

Beehler, B. M. (ed.). 1993. Papua New Guinea conservation needs assessment. Volume 2. Biodiversity Support Program, Washington, D.C. 434 pp.

Beehler, B. M., C. G. Burg, C. Filardi, and K. Merg. 1994. Birds of the Lakekamu-Kunimaipa Basin. Muruk 6: 1-8.

Beehler, B. M., J. B. Sengo, C. Filardi, and K. Merg. 1995. Documenting the lowland rainforest avifauna in Papua New Guinea—effects of patchy distributions, survey effort and methodology. Emu 95: 149-162.

Beehler, B. M. and R. Bino. 1995. Yellow-eyed Starling *Aplonis mystacea* in Central Province, Papua New Guinea. Emu 95: 68-71.

Bell, H. A. 1982. A bird community of lowland rainforest in New Guinea. I. Composition and density of the avifauna. Emu 82: 24-41.

Bennett, D. 1996. Warane der Welt, Welt der Warane. Edition Chimaira, Frankfurt am Main. 383 pp.

Bibby, C. J., N. D. Burgess and D. A. Hill. 1992. Bird census techniques. Academic Press, London. 257 pp.

Bolton, B. 1977. The ant tribe Tetramoriini (Hymenoptera: Formicidae). The genus *Tetramorium* Mayr in the Oriental and Indo-Australian regions, and in Australia. Bulletin of the British Museum of Natural History (Entomology) 36: 67-151.

Bolton, B. 1995. A new general catalogue of the ants of the world. Harvard University Press, Cambridge, MA. 504 pp.

Brown, H A 1973. The Eleman language family. *In* Franklin, K. J. (ed.). The linguistic situation in the Gulf District and adjacent area, Papua New Guinea, C26. Pacific Linguistics, Canberra. 597 pp.

Brown, W. L., Jr. 1958. Predation of arthropod eggs by the ant genera *Proceratium* and *Discothyrea*. Psyche 64: 115.

Campion, H. 1915. Report on the odonata collected by the British Ornithologist's Union Expedition and the Wollaston Expedition in Dutch New Guinea. Transactions of the Zoological Society of London 20: 485-492.

Cheesman, L. E. 1951. A collection of *Polistes* from Papuasia in the British Museum. Annual Magazine of Natural History (12) 4:982-993.

Cheesman, L. E. 1952. *Ropalidia* of Papuasia. Annual Magazine of Natural History (12) 5:1-26.

Cogger, H. G. 1992. Reptiles and amphibians of Australia. Fifth ed. Cornell University Press, Ithaca. 775 pp.

Coode, M. J. E. 1978. Combretaceae. *In* Womersley, J. (ed.) Handbooks of the flora of Papua New Guinea 1: 43-110.

Coode, M. J. E. 1981. Elaeocarpaceae. *In* Henty, E. E. (ed.) Handbooks of the flora of Papua New Guinea 2: 38-185.

Copeland, E. B. 1949. Aspleniaceae and Blechnaceae of New Guinea. Philippine Journal of Science 78: 207-229.

Cowie, A. T. 1984. Lactation. *In* Austin, C. R. and R. V. Short (eds.). Reproduction in Mammals: 3. Hormonal Control of Reproduction. Cambridge University Press, Cambridge. 244 pp.

Davidson, D. W. 1997. The role of resource imbalances in the evolutionary ecology of tropical arboreal ants. Biological Journal of the Linnean Society 61: 153-182.

Davies, H. L., R. D. Winn, and P. KenGemar. 1996. Evolution of the Papuan Basin - a view from the orogen. *In* Buchanan, P. G. (ed.). Petroleum exploration, development and production in Papua New Guinea. Proceedings of the Third Papua New Guinea Petroleum Convention. Pp. 53-62. PNG Chamber of Mines and Petroleum, Port Moresby.

Davis, S. D. 1995a. Regional overview: south east Asia (Malesia). *In* Davis, S. D., V. H. Heywood and A. C. Hamilton (eds.). Centres of Plant Diversity. Vol 2 Asia, Australasia and the Pacific. Pp. 229-432. World Wildlife Fund and IUCN- The World Conservation Union, Cambridge.

Davis, S. D. 1995b. Identifying sites of global importance for conservation: the IUCN/WWF Centres of Plant Diversity Project. *In* Primack, R. B. and T. E. Lovejoy (eds.). Ecology, Conservation, and Management of Southeast Asian Rainforests. Pp. 176-203. Yale University Press, New Haven.

Dawkins, H. C. 1959. The volume increment of natural tropical high forest and limitations on its improvement. Empire Forestry Review 38: 175-180.

Department of Environment and Conservation. 1996. Forest types susceptible to conventional logging practices. Discussion of forest inventory mapping program (FIM) vegetation types. Draft policy document distributed for review.

de Rooij, N. 1915. The reptiles of the Indo-Australian Archipelago. I. Lacertilia, Chelonia, Emydosauria. E. J. Brill, Leiden. 384 pp.

Donnellan, S. C., and K. P. Aplin. 1989. Resolution of cryptic species in the New Guinea lizard *Sphenomorphus jobiensis* (Scincidae) by electrophoresis. Copeia 1989: 81-88.

Dow, D. B., J. A. J. Smit and R. W. Page. 1974. Wau, Papua New Guinea. 1:250,000 Geological Series- Explanatory Notes. Department of Minerals and Energy. Australian Government Publishing Service, Canberra.

Filer, C. 1994. The nature of the human threat to Papua New Guinea's biodiversity endowment. *In* Sekhran, N and S. Miller (eds.). 1994 Papua New Guinea Country Study on Biological Diversity. A report to the United Nations Environment Program, Waigani, Papua New Guinea. Department of Environment and Conservation, Conservation Resource Centre; and Nairobi, Kenya, Africa Centre for Resources and Environment (ACRE). 438 pp.

Filer, C. 1997. Compensation, rent and power in Papua New Guinea. *In* Toft, S. (ed.). Compensation for Resource Development in Papua New Guinea. Papua New Guinea Law Reform Commission Monograph No. 6: 156-189. Resource Management in Asia Pacific, and National Centre for Development Studies, Pacific Policy Paper #24, Port Moresby and Canberra. 201 pp.

Filer, C S and W. Iamo. 1989. Base-Line Planning Study for the Lakekamu Gold Project, Gulf Province. Mimeo, Department of Anthropology and Sociology, University of Papua New Guinea January (revised draft). 96 pp.

Fittkau, E. J., and H. Klinge. 1973. On biomass and trophic structure of the central Amazonian rain forest ecosystem. Biotropica 5: 2-14.

Flannery, T. 1995. Mammals of New Guinea. Reed Books, NSW, Australia and Cornell University Press, Ithaca. 568 pp.

Fleming, T. H. 1988. The Short-tailed Fruit Bat: A Study in Plant-Animal Interactions. The University of Chicago Press, Chicago. 365 pp.

Forster, P. I. 1990. Notes on Asclepiadaceae, 2. Austobaileya 3: 273-289.

Forster, P. I. 1991. *Cryptolepis lancifolia* (Asclepiadaceae: Periplocoideae), a new species from Irian Jaya. Blumea 35: 381-382.

Forster, P. I. 1993. Conspectus of *Cryptolepis* R. Br. (Asclepiadaceae: Periplocoideae) in Malesia. Austrobaileya 4: 67-73.

Georges, A. and M. Adams. 1996. Electrophoretic delineation of species boundaries within the short-necked freshwater turtles of Australia (Testudines: Chelidae). Zoological Journal of the Linnean Society 118: 241-260.

Gentry, A. H. 1988. Tree species richness of upper Amazonian forests. Proceedings of the National Academy of Science (USA) 85: 156-159.

Gewertz, D. and F. Errington. 1991. Twisted histories, altered contexts: representing the Chambri in a world system. Cambridge University Press, New York. 264 pp.

Girard, C. F. 1858. Herpetology. United States Exploring Expedition during the years 1838, 1839, 1840, 1841, 1842 under the command of Charles Wilkes, U. S. N. Vol. 20. J. B. Lippincott & Co., Philadelphia. 496 pp.

Glucksman, J., G. West and T. M. Berra. 1976. The introduced fishes of Papua New Guinea with special reference to *Tilapia mossambica*. Biological Conservation 9: 37-44.

Gressitt, J. L. 1959. Wallace's line and insect distribution. Proceedings of the XVth International Congress of Zoology (1958), pp. 66-68.

Gressitt, J. L. 1961. Problems in the zoogeography of Pacific and Antarctic insects. Pacific Insect Monographs 2: 1-94.

Günther, A. 1877. Descriptions of three new species of lizards from islands of the Torres Strait. Annual Magazine of Natural History 19: 413-415.

Haines, A. K. 1979. Purari River (Wabo) hydroelectric scheme environmental studies Vol. 6 - An ecological survey of fish of the lower Purari River system, Papua New Guinea. Technical report printed by Office of Environment and Conservation and Department of Minerals and Energy, Port Moresby. 101 pp.

Hallpike, C R. 1977. Bloodshed and vengeance in the Papua mountains. Oxford University Press, New York. 317 pp.

Hammermaster, E. T. and J. C. Saunders. 1995. Forest Resources and Vegetation Mapping of Papua New Guinea. PNGRIS Publ. no. 4. CSIRO and AIDAB, Canberra.

Hartley, H. G. and L. M. Perry. 1973. A provisional key and enumeration of species of *Syzygium* (Myrtaceae) from Papuasia. Journal of the Arnold Arboretum 54: 160-227.

Hau'ofa, E. 1981. Mekeo: inequality and ambivalence in a village society. Australian National University Press, Canberra. 339 pp.

Hesse, P. R. 1972. A Textbook of Soil Chemical Analysis. Chemical Publication Co., New York.

Heyer, W. R., M. A. Donnelly, R. W. McDiarmid, L. C. Hayek, and M. S. Foster (eds.). 1994. Measuring and Monitoring Biological Diversity - Standard Methods for Amphibians. Smithsonian Institute Press. Washington, D.C. 364 pp.

Höft, R. 1992. Plants of New Guinea and the Solomon Islands. Wau Ecology Institute Handbook No. 13, Wau. 168 pp.

Hölldobler, B., and E. O. Wilson. 1990. The ants. Belknap Press, Harvard University, Cambridge, MA. 732 pp.

Hölldobler, B., and E. O. Wilson. 1994. Journey to the ants. Belknap Press, Harvard University, Cambridge, MA. 228 pp.

Holtum, R. E. 1954. A revised flora of Malaya, volume II, ferns of Malaya. Government Printing Office, Singapore. 627 pp.

Holtum, R. E. 1986. Studies in the fern genera allied to *Tectaria* Cav. VI: a conspectus of genera in the Old World regarded as related to *Tectaria*, with descriptions of two genera. Gardens Bulletin (Singapore) 39: 153-167.

Holtum, R. E. 1991. *Tectaria* group. Flora Malesiana ser II. 2(1): 1-132.

Hou, D., K. Larsen and S. S. Larsen. 1996. Caesalpinaceae (Leguminosae-Caesalpinioideae). Flora Malesiana ser I 12(2): 409-730.

Hubbell, S. P. and R. B. Foster. 1992. Short-term dynamics of a neotropical forest: why ecological research matters to tropical conservation and management. Oikos 63: 48-61.

Ingram, G. J. and J. A. Covacevich. 1988. Revision of the genus *Lygisaurus* De Vis (Scincidae: Reptilia) in Australia. Memoirs of the Queensland Museum 25: 335-354.

IUCN. 1996. 1996 IUCN Red List of Threatened Animals. IUCN, Gland. 368 pp.

Johnston, G. R., and S. J. Richards. 1993. Observations on the breeding biology of a microhylid frog (Genus *Oreophryne*) from New Guinea. Transactions of the Royal Society of South Australia 117: 105-107.

Johns, R. J. 1986. The instability of the tropical ecosystem in New Guinea. Blumea 31: 341-371.

Kiapranis, R. 1991. Plant species enumeration in a lowland rain forest in Papua New Guinea. *In* Taylor, D. A. and K. G. MacDicken (eds.). Research on multipurpose tree species in Asia. Pp. 98-101. Winrock International Institute for Agricultureal Development, Bangkok.

King, M., and P. Horner. 1989. Karyotypic evolution in *Gehyra* (Gekkonidae: Reptilia). V. A new species from Papua New Guinea and morphometrics of *Gehyra oceanica* (Lesson). Beagle 6: 169-178.

Kirsch, S. 1997. Regional dynamics and conservation in Papua New Guinea: The Lakekamu River Basin project. The Contemporary Pacific 9: 97-121.

Kostermans, A. J. G. H. 1962. The genera *Belotia* Rich. and *Trichospermum* Bl. (Tiliaceae). Reinwardtia 6: 277-279.

Linnekin, J. 1992. On the theory and politics of cultural construction in the Pacific. Oceania 62: 249-269.

Lovejoy, T. E., R. O. Bierregaard Jr., A. B. Rylands, J. R. Malcolm, C. E. Quintela, L. H. Harper, K. S. Brown Jr., A. H. Powell, G. V. N. Powell, H. O. R. Schubart, and M. B. Hay. 1986. Edge and other effects of isolation on Amazon forest fragments. *In* Soule, M. E. (ed.). Conservation biology: the science of scarcity and diversity. Pp. 257-285. Sinauer, Sunderland.

Loveridge, A. 1948. New Guinean reptiles and amphibians in the Museum of Comparative Zoology and the United States National Museum. Bulletin of the Museum of Comparative Zoology 101: 305-430.

Mabberley, D. J., C. M. Pannell and A. M. Sing. 1995. Meliaceae. Flora Malesiana ser. I 12 (1): 1-388

Mack, A. L. and D. D. Wright. 1996. Notes on occurrence and feeding of birds at Crater Mountain Biological Research Station, Papua New Guinea. Emu 96: 89-101.

Mack, A. L. and D. D. Wright. 1998. The vulturine parrot, *Psittrichas fulgidus*, a threatened New Guinea endemic: notes on its biology and conservation. Bird Conservation International in press.

McArthur, M. 1971. Men and spirits in the Kunimaipa valley. *In* Hiatt, L. R. and C. Jayawardena (eds.). Anthropology in Oceania: essays presented to Ian Hogbin. Chandler, San Francisco.

McCoy, M. 1980. Reptiles of the Solomon Islands. Wau Ecology Institute Handbook No. 7, Wau. 82 pp.

McDowell, S. B. 1979. A catalogue of the snakes of New Guinea and the Solomons, with special reference to those in the Bernice P. Bishop Museum. Part III. Boinae and Acrochordoidea. Journal of Herpetology 13: 1-92.

Menzies, J. I. 1973. Handbook of common New Guinea frogs. Wau Ecology Institute Handbook No. 1, Wau. 75 pp.

Menzies, J. I. 1987. A taxonomic revision of the Papuan *Rana* (Amphibia: Ranidae). Australian Journal of Zoology 35: 373-418.

Merrill, E. D. and L. M. Perry. 1943. Plantae Papuanae Archboldianae, XIII. Journal of the Arnold Arboretum 24: 422-439.

Michener, C. D. 1965. A classification of the bees of the Australian and South Pacific Regions. Bulletin of the American Museum of Natural History 130: 1-362.

Mori, S. A., B. M. Boom, A. M. de Carvalho and T. S. Dos Santos. 1983. Southern Bahian moist forests. Botanical Review 49: 155-232.

Mosko, M. 1985. Quadripartite structures: categories, relations and homologies in bush Mekeo culture. Newridge University Press, New York. 298 pp.

Munro, I. S. R. 1958. The fishes of the New Guinea region. Papua New Guinea Agriculture Journal 10: 97-369.

Mys, B. 1988. The zoogeography of the scincid lizards from the north Papua New Guinea (Reptilia: Scincidae). I. The distribution of the species. Bulletin Institut Royal des Sciences Naturelles Belgique Biologie 58: 127-184.

Nadarajah, T. 1993. Forest policy and conservation of Papua New Guinea's forests: a critique. National Research Institute Discussion Paper no 70. The National Research Institute, Boroko.

Ngan, Phung Trung. 1965. A revision of the genus *Wrightia* (Apocynaceae). Annals of the Missouri Botanical Garden 52: 114-175.

Oatham, M. and B. M. Beehler. 1997. Richness, taxonomic composition, and species patchiness in three lowland forest tree plots in Papua New Guinea. *In* Proceedings of the international symposium for measuring and monitoring forests and biological diversity; the international networks of biodiversity plots. Smithsonian Institute/Man and Biosphere Biodiversity Program (SI/MAB).

Orsak, L. 1991. The Wau Ecology Institute's 'Insect Ranch' Programme. Unpublished pamphlet, Wau Ecology Institute. 7 pp.

Osborne, P. L. 1995. Biological and cultural diversity in Papua New Guinea: conservation, conflicts, constraints and compromise. Ambio 24:231-237.

O'Shea, M. T. 1990. The highly and potentially dangerous elapids of Papua New Guinea. In Gopalakrishnakone, P. and L. M. Chou (eds.). Snakes of medical importance in the Asian-Pacific region. pp. 585-640. Venom and Toxin Research Group, National University of Singapore, Singapore.

O'Shea, M. T. 1996. A guide to the snakes of Papua New Guinea. Independent Publishing, Independent Group PTY, Port Moresby. 239 pp.

Page, A. L., R. H. Miller and D. R. Keeney. 1982 Methods of Soil Analysis. Part 2. Chemical and Microbiological properties. 2nd edition. American Society of Agronomy, Inc. and Soils Science Society of America, Inc., Madison.

Paijmans, K. 1970. An analysis of four tropical rain forests in Papua New Guinea. Journal of Ecology 58: 77-101.

Papua New Guinea National Statistical Office. 1993. 1990 National Population Census. Port Moresby: National Statistical Office.

Parenti, L. R., and G. R. Allen. 1991. Fishes of the Gogol River and other coastal habitats, Madang Province, Papua New Guinea. Ichthyological Explorations of Freshwaters 1: 307-320.

Peters, W., and G. Doria. 1878. Catalago dei Retilli e dei Batraci racolti da O. Beccari, L.M. D'Albertis e A.A. Bruijn nella Sotto-Regione Austro-Malese. Annali del Museo Civico di Storia Naturale di Genova 13: 323-450.

Philipson, W. R. 1980. A revision of Levieria (Monimiaceae). Blumea 26: 373-385.

Philipson, W. R. 1986. Monimiaceae. Flora Malesiana, Ser I. 10(2): 255-326.

Phillips, O. L., P. Hall, A. H. Gentry, S. A. Sawyer and R. Vasquez. 1994. Dynamics and species richness of tropical rain forests. Proceedings of the National Academy of Sciences (USA) 91: 2805-2809.

Polhemus, D. A. 1995. A preliminary biodiversity survey of aquatic heteroptera and other aquatic insect groups in the Kikori River Basin, Papua New Guinea *In* Field Survey of Biodiversity in the Kikori River Basin, Papua New Guinea. World Wildlife Fund, Washington.

Polhemus, D. A. 1997. Unpublished report to Freeport Mining Company.

Proctor, J., J. M. Anderson, P. Chai and H. W. Vallack. 1983. Ecological studies in four contrating rain forests in Gunung Mulu National Park, Sarawak. Journal of Ecology 71: 237-260.

Rhodin, A. G. J., R. A. Mittermeier, and P. M. Hall. 1993. Distribution, osteology, and natural history of the Asian giant softshell turtle, *Pelochelys bibroni*, in Papua New Guinea. Chelonian Conservation and Biology 1(1):19-30.

Richards, O. W. 1978. The Australian social wasps (Hymenoptera: Vespidae). Australian Journal of Zoology Supplemental Series No. 61: 1-132.

Richards, S. J. 1992. The tadpole of the Australopapuan frog *Rana daemeli*. Memoirs of the Queensland Museum 32: 138.

Richards, S. J., K. R. McDonald and R. A. Alford. 1993. Declines in populations of Australia's endemic tropical rainforest frogs. Pacific Conservation Biology 1: 66-77.

Ris, F. 1913. Die Odonata von Dr. H. A. Lorentz' expedition nach sudwest-neu-guinea 1909 und einige Odonata von Waigeu. Nova Guinea (Zoologie) 9: 471-512.

Roberts, T. R. 1978. An ichthyological survey of the Fly River in Papua New Guinea with descriptions of new species. Smithsonian Contributions in Zoology No. 281: 1-72.

Ruddle, K., and East-West Technology and Development Institute. 1982. Palm Sago: A Tropical Starch from Marginal Lands. Honolulu: East-West Center. 202 pp.

Safford, R. 1996. A nesting colony of Yellow-eyed Starlings *Aplonis mystacea*. Emu 96: 140-142.

Schieffelin, E. L. 1976. The Sorrow of the Lonely and the Burning of the Dancers. St. Martin's Press, New York. 243 pp.

Schwarz, H. F. 1939. The Indo-Malayan species of *Trigona*. Bulletin of the American Museum of Natural History 76: 83-141.

Sekhran, N and S. Miller (eds.). 1994 Papua New Guinea Country Study on Biological Diversity. A report to the United Nations Environment Program, Waigani, Papua New Guinea. Department of Environment and Conservation, Conservation Resource Centre; and Nairobi, Kenya, Africa Centre for Resources and Environment (ACRE). 438 pp.

Shattuck, S. O. 1992. Generic revision of the ant subfamily Dolichoderinae (Hymenoptera: Formicidae). Sociobiology 21: 1-181.

Sinclair, J. 1955. A revision of the Malayan Annonaceae. Gardens Bulletin (Singapore) 14: 149-516.

Sleumer, H. 1986. A revision of the genus *Rapanea* Aubl. (Myrsinaceae) in New Guinea. Blumea 31: 245-269.

Snelling, R. R. 1981. Systematics of social Hymenoptera, *In* Hermann, H. (ed.). Social Insects pp. 369-453. Academic Press, New York.

Soepadmo, E. 1995. Plant diversity of the Malesian tropical rainforest and its phytogeographical and economic significance. Pp. 19-40. *In* Primack, R. B. and T. E. Lovejoy (eds.). Ecology, Conservation, and Management of Southeast Asian Rainforests. Pp. 19-40. Yale University Press, New Haven.

Sohmer, S. H., R. Kiapranis, A. Allison and W. Takeuchi. 1991. Report on the Hunstein River expedition- 1989. Unpublished manuscript. 70 pp.

Sokal, R. R. and F. J. Rohlf. 1995. Biometry: the principles and practice of statistics in biological research. 3rd ed. W. H. Freeman and Co., New York.

Spradbery, J. P. 1973. Wasps: an account of the biology and natural history of solitary and social wasps. University of Washington Press, Seattle. 408 pp.

Stephen, M. 1995. A'isa's Gifts. University of California Press, Berkeley. 381 pp.

Starr, C. K. 1992. The social wasps (Hymenoptera: Vespidae) of Taiwan. Bulletin of the National Museum of Natural Science (Taiwan) 3: 93-138.

Takeuchi, W. 1995. A preliminary assessment of the Waskuk flora. Unpublished report submitted to the PNG Forest Research Institute. 136 pp.

Taylor, R. W. 1965. A monographic revision of the rare tropicopolitan ant genus *Probolomyrmex* Mayr (Hymenoptera: Formicidae). Transactions of the Royal Entomological Society (London) 117: 345-365.

Taylor, R. W. 1967. A monographic revision of the ant genus *Ponera* Latreille (Hymenoptera: Formicidae). Pacific Insects Monograph 13. 112 pp.

Turner, I. M. and R. T. Corlett. 1996 The conservation value of small, isolated fragments of low-land tropical rain forest. Trends in Ecology and Evolution 11: 330-333.

van Heusden, E. C. H. 1992. Flowers of Annonaceae: morphology, classification, and evolution. Blumea Supplement 7. 218 pp.

Vecht, J. van der. 1957. The Vespinae of the Indo-Malayan and Papuan areas (Hymenoptera, Vespidae). Zooloogische Verhandlingen 34: 1-83.

Vecht, J. van der. 1966. The East-Asiatic and Indo-Australian species of Polybioides Buysson and Parapolybia Saussure (Hym., Vespidae). Zooloogische Verhandlingen 82:1-42.

Vecht, J. van der. 1971. The subgenera *Megapolistes* and *Stenopolistes* in the Solomon Islands (Hymenoptera, Vespidae, Polistes Latreille). Entomology Essays Commemorating Retired Professor K. Yasumatsu, pp. 87-106.

Vitousek, P. M. 1984. Litterfall, nutrient cycling, and nutrient limitation in tropical forests. Ecology 65: 285-298.

Webb, R. G. 1995. Redescription and neotype designation of *Pelochelys bibroni* from southern New Guinea (Testudines: Trionychidae). Chelonian Conservation and Biology 4: 301-310.

Weber, M. 1913. Susswasserfische aus Niederlandissh Sud-und Nord Neu Guinea. Nova Guinea (Leiden). 9(4): 513-613.

Whitaker, R. and Z. Whitaker. 1982. Reptiles of Papua New Guinea. Wildlife in Papua New Guinea. Vol. 82/2. Hebamo Press, Boroko. 53 pp.

Willey, R. B., and W. L. Brown, Jr. 1983. New species of the ant genus *Myopias* (Hymenoptera: Formicidae: Ponerinae). Psyche 90: 249-285.

Wilson, E. O. 1958a. Studies on the ant fauna of Melanesia. I. The tribe Leptogenyini. II. The tribes Amblyoponini and Platythyreini. Bulletin of the Museum of Comparative Zoology 118: 101-153.

Wilson, E. O. 1958b. Studies on the ant fauna of Melanesia. III. *Rhytidoponera* in western Melanesia and the Moluccas. IV. The tribe Ponerini. Bulletin of the Museum of Comparative Zoology 119: 303-371.

Wilson, E. O. 1959a. Studies on the ant fauna of Melanesia. V. The tribe Odontomachini. Bulletin of the Museum of Comparative Zoology 120: 483-510.

Wilson, E. O. 1959b. Studies on the ant fauna of Melanesia. VI. The tribe Cerapachyini. Pacific Insects 1: 39-57.

Wilson, E. O. 1959c. Some ecological characteristics of ants in New Guinea rain forests. Ecology 40: 437-447.

Wilson, E. O. 1964. The true army ants of the Indo-Australian area (Hymenoptera: Formicidae: Dorylinae). Pacific Insects 6: 427-483.

Wilson, E. O. 1987. The arboreal ant fauna of Peruvian Amazon forests: a first assessment. Biotropica 19: 245-251.

Wilson, E. O., and R. W. Taylor. 1967. The ants of Polynesia. Pacific Insects Monograph No. 14. 109 pp.

Woodruff, D. S. 1972. Amphibians and reptiles from Simbai, Bismarck-Schrader Range, New Guinea. Memoirs of the National Museum of Victoria 33: 57-64.

Wright, D. D., J. H. Jessen, P. Burke and H. G. S. Garza. 1997. Tree and liana enumeration and diversity on a one-hectare plot in Papua New Guinea. Biotropica 29: 250-260.

Zar, J. H. 1984. Biostatistical Analysis. 2nd Ed. Prentice Hall, Englewood Cliffs.

Zug, G. R. and B. R. Moon. 1995. Systematics of the Pacific slender-toed geckos, Nactus pelagicus complex: Oceania, Vanuatu, and Solomon Island populations. Herpetologica 51: 77-90.

APPENDICES

Species Accumulation Curves

Species accumulation curves (or species/effort curves) for several taxa surveyed on the Lakekamu RAP survey 1996. Note that a curve that levels off (asymptotes) indicates that the sampling method in use revealed most of the species present at the site that can be revealed *using that method*. A level curve does not mean that all species present had been revealed, only that additional sampling with that method would not reveal many more species. A change in methodology (for example moving mist nets into the canopy to sample birds or bats) could reveal more species. The curves below generally indicate that sampling was fairly thorough for the particular methods employed despite the relative short sampling period of the RAP. See each chapter for specific details.

Appendix 1A   Species area curves for five one-hectare plots in PNG. The two dark symbols represent data from this study, the Nagore and Si plots are also within the Lakekamu Basin (Oatham and Beehler 1997) and the CMWMA plot is in upper hill-lower montane forest in Simbu Province (Wright *et al.* 1997). Points represent the cumulative number of species in subplots constituting each one hectare plot.

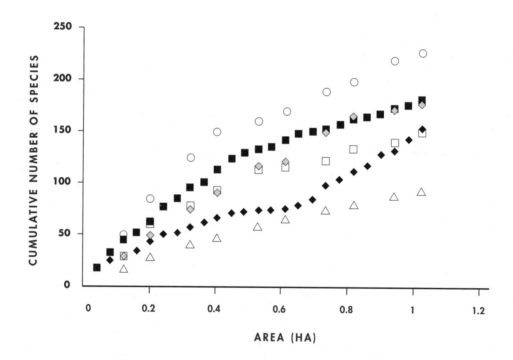

- ◆ Hill Plot
- ■ Alluvial Plot
- △ Si River
- ◇ Nagore S
- □ Nagore N
- ○ CMWMA

Appendix 1B. Species accumulation curve for ants surveyed. Sample units are field days by the entomologist Roy Snelling. Although each day is not a standard, invariable unit, the shape of the curve does indicate the techniques employed continued to add new species up until the last day. Most species these techniques could be expected to reveal had been collected within 40 days.

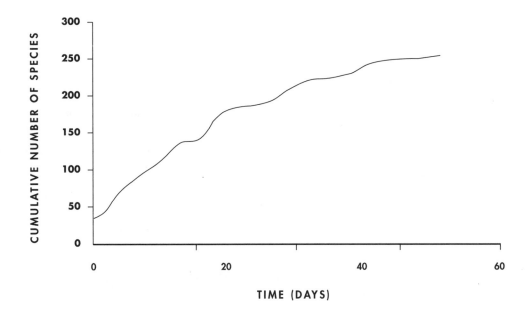

Appendix 1C   Species-effort curve for herpetofauna censused using the 5 m x 5 m leaf litter plot methodology.

SPECIES ACCUMULATION - LEAF LITTER PLOT

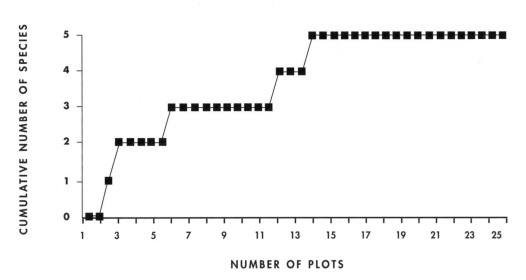

Appendix 1D   Species-effort curve for the herpetofauna censused using Visual Encounter Survey (VES) transects.

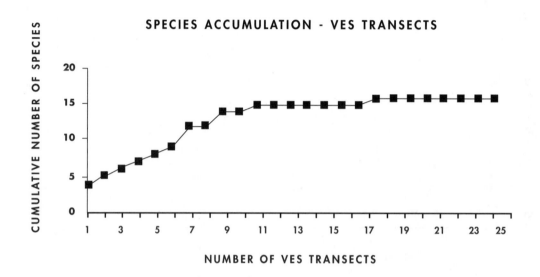

Appendix 1E   Species accumulation curve for birds at the Ivimka study area. Species recorded by all means available (netted, observed, heard) by all observers pooled.

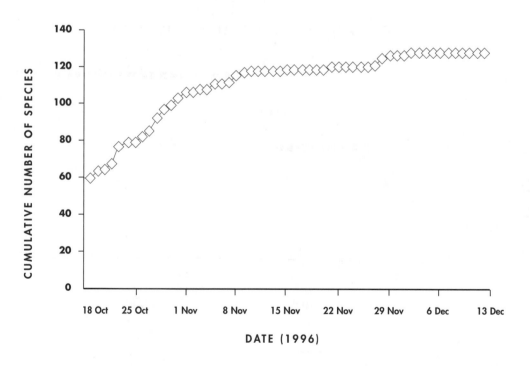

Appendix 1F Species accumulation curve for avian point counts. Point counts were ten minutes at each station; stations 200 m apart. The total number of species recorded does not include individuals heard but not positively identified to species.

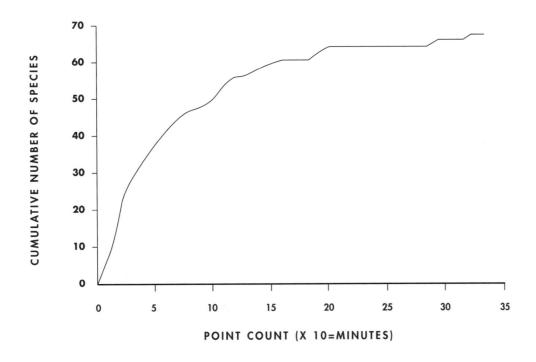

POINT COUNT (X 10=MINUTES)

Appendix 1G Species accumulation curve for mist net captures of birds around the Ivimka Research Station. Nets were run 29 days between 18 October to 3 December 1996. Net meter hours is unit that standardizes for size of nets and amount of time they are open.

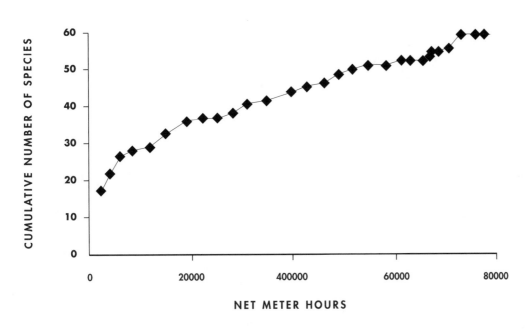

NET METER HOURS

Appendix 1H   The cumulative number of species captured as the number of trap nights increased during the study period. Each arrow depicts the last night before the trap line location was changed. The first four trap lines were in alluvial forest and the last two lines were in hill forest.

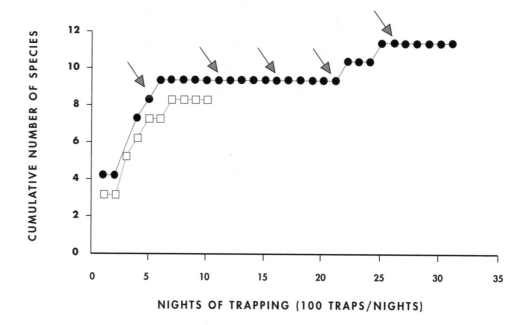

Appendix 1I   The cumulative number of bat species captured as the number of mistnet-meter-hours increased during the study period. Each arrow depicts the last night before the mistnet lane location was changed.

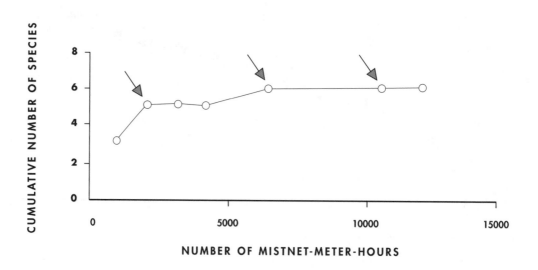

Tree Species Recorded on Two One-Hectare Vegetation Plots

Species of trees recorded on the two one-hectare plots, one in hill forest and one in alluvial forest at the IRS. All trees ≥ 10 cm DBH are included in the list. Nomenclature follows Höft (1992).

FAMILY	GENUS	SPECIES	ALLUVIAL	HILL
Actinidiaceae	*Saurauia*	sp.		x
Anacardiaceae	*Buchanania*	*arborescens*	x	
	Campnosperma	*brevipetiolatum*	x	
	Dracontomelon	*dao*		x
Annonaceae	*Cananga*	*odorata*	x	x
	Cyathocalyx	aff. *petiolatus*		x
	Goniothalamus	sp.		x
	Miliusa	sp.	x	
	Polyalthia	*oblongifolia*	x	
	Popowia	*pisocarpa*	x	
	Xylopia	aff. *caudata*		x
	Xylopia	*papuana*	x	
Apocynaceae	*Cerbera*	aff. *floribunda*	x	
	Ochrosia	*ficilifolia*	x	
	Tabernaemontana	*pandacaqui*	x	x
	Wrigthia	aff. *laevis*	x	x
	Wrigthia	sp.	x	
Arecaceae	*Gulubia*	sp.	x	x
	Orania	sp.	x	x
Barringtoniaceae	*Barringtonia*	*novae-hyberniae*	x	
	Barringtonia	sp.	x	
Burseraceae	*Canarium*	*indicum*		x
	Canarium	*maluensis*	x	
	Haplolobus	*floribundus floribundus*	x	x
	Haplolobus	*floribundus solomonensis*	x	
	Haplolobus	*furfuraceaous*	x	x
	Haplolobus	*lanceolatus*	x	
	Haplolobus	*pubescens*	x	
	Haplolobus	sp.	x	
Caesalpiniaceae	*Crudia*	*papuana*	x	

FAMILY	GENUS	SPECIES	ALLUVIAL	HILL
	Maniltoa	sp. A	x	x
	Maniltoa	sp. B	x	
Celastraceae	*Lophopetalum*	*torricelense*	x	
Chrysobalanaceae	*Cyclandrophora*	*laurina*	x	x
	Parastemon	*versteeghii*	x	x
Clusiaceae	*Calophyllum*	*papuanum*		x
	Calophyllum	*soulattri*	x	x
	Garcinia	*celebica*	x	
	Garcinia	*hunsteinii*	x	x
	Garcinia	aff. *ledermanii*	x	
	Garcinia	*maluensis*	x	x
	Garcinia	sp.		x
	Garcinia	sp. nov.		x
	Garcinia	*warrenii*		x
Combretaceae	*Terminalia*	aff. *complanata*	x	
	Terminalia	aff. *kaernbachii*		x
	Terminalia	sp.	x	
Cunoniaceae	*Caldcluvia*	*nymanii*	x	
	Caldcluvia	sp.	x	
Dipterocarpaceae	*Anisoptera*	*polyandra*	x	x
Ebenaceae	*Diospyros*	*maritima*	x	
	Diospyros	aff. *villosluscula*	x	
	Diospyros	sp.	x	
Elaeocarpaceae	*Elaeocarpus*	aff. *coloides*	x	
	Elaeocarpus	aff. *densiflorus*	x	
	Elaeocarpus	*culminicola*	x	
	Elaeocarpus	*schoddei*	x	
	Elaeocarpus	*sepikanus*	x	
	Elaeocarpus	*womersleyi*	x	
	Elaeocarpus	sp. A	x	
	Elaeocarpus	sp. B	x	
	Sloanea	*nymanii*	x	x
	Sloanea	sp.	x	

FAMILY	GENUS	SPECIES	ALLUVIAL	HILL
Erythroxylaceae	*Erythroxylum*	*ecarinathum*	x	
Euphorbiaceae	*Antidesma*	*moluccanum*	x	
	Aporusa	*papuana*	x	x
	Baccaurea	*philippinensis*	x	
	Baccaurea	sp.	x	
	Baccaurea	*tristis*	x	
	Blumeodendron	*papuanum*	x	
	Breynia	*cernua*		x
	Bridelia	*penangiana*	x	
	Cleistanthus	aff. *insignis*	x	
	Croton	sp.	x	
	Drypetes	cf. *bordenii*	x	x
	Endospermum	*medullosum*	x	
	Fontainea	*pancheri*		x
	Macaranga	*densiflora*	x	
	Macaranga	*fimbriata*	x	x
	Macaranga	*punctata*	x	
	Mallotus	*echinatus*		x
	Neoscortechinia	*forbesii*	x	
	Pimelodendron	*amboinicum*		x
Fagaceae	*Lithocarpus*	aff. *rufovillosus*		x
	Lithocarpus	aff. *vinkii*		x
	Lithocarpus	*celebicus*	x	x
	Lithocarpus	sp.	x	
Flacourtiaceae	*Pangium*	*edule*	x	
	Ryparosa	*javanica*	x	x
	Trichadenia	*philippinensis*	x	x
Gnetaceae	*Gnetum*	*gnemon*	x	x
Grossulariaceae	*Polyosma*	*integrifolia*	x	x
	Polyosma	sp.	x	x
Icacinaceae	*Gomphandra*	*montana*	x	
	Gonocaryum	*littorale*	x	
	Platea	*excelsa*	x	x

FAMILY	GENUS	SPECIES	ALLUVIAL	HILL
Lauraceae	*Actinodaphne*	*nitida*		x
	Beilschmiedia	*acutifolia*	x	
	Beilschmiedia	aff. *bangkae exudens*		x
	Beilschmiedia	*bangkae*	x	x
	Beilschmiedia	sp.	x	
	Cryptocarya	aff. *multipaniculata*	x	
	Cryptocarya	aff. *renicarpa*	x	x
	Cryptocarya	*caudata*		x
	Cryptocarya	cf. *massoy*		x
	Cryptocarya	cf. *rigida*	x	
	Cryptocarya	*crassinervia*	x	x
	Cryptocarya	*eugenioides*		x
	Cryptocarya	*idenburgensis*		x
	Cryptocarya	*multinervis*		x
	Cryptocarya	sp. A		x
	Cryptocarya	sp. B		x
	Cryptocarya	sp. C	x	x
	Cryptocarya	*maluense*	x	
	Dehaasia	sp.	x	
	Endiandra	*brassii*	x	x
	Endiandra	*clemensii*		x
	Endiandra	sp.	x	x
	Litsea	*collina*		x
	Litsea	*firma*	x	x
	Litsea	sp.	x	x
	Litsea	*timoriana*	x	x
	Phoebe	*forbessii*		x
Loganiaceae	*Fagraea*	aff. *berteroana*		x
	Fagraea	*racemosa*	x	
	Fagraea	sp.	x	
	Geniostoma	sp.	x	
	Neuburgia	*corynocarpa*	x	
	Strychnos	*minor*		x

FAMILY	GENUS	SPECIES	ALLUVIAL	HILL
Magnoliaceae	*Elmerrillia*	*papuana*	X	X
Melastomataceae	*Memecylon*	*schraderbergense*		X
Meliaceae	*Aglaia*	aff. *ganggo (=A. silvestris)*		X
	Aglaia	*goebeliana (= A. rimosa)*		X
	Aglaia	*argenta*	X	X
	Aglaia	*cucullata*		X
	Aglaia	*sapindina*		X
	Chisocheton	*weinlandii*		X
	Chisocheton	sp.	X	
	Dysoxylum	aff. *kaniense*	X	X
	Dysoxylum	*alliaceum*	X	X
	Dysoxylum	*archboldianum*	X	X
	Dysoxylum	*gaudichaudianum*	X	X
	Dysoxylum	sp.	X	X
Mimosaceae	*Abarema*	cf. *kiahii*		X
	Serianthes	*minahassae*	X	
Monimiaceae	*Levieria*	aff. *montana*		X
	Levieria	*beccariana*	X	
	Steganthera	*hirsuta*		X
	Steganthera	sp.		X
Moraceae	*Artocarpus*	aff. *sepicanus*	X	
	Artocarpus	*fretessii*	X	
	Ficus	*mollior*	X	
	Ficus	aff. *nodosa*	X	
	Ficus	*odoardi*	X	
	Ficus	*variegate*	X	X
	Parartocarpus	*venenosus*	X	X
	Streblus	*glaber*		X
Myristicaceae	*Gymnacranthera*	*paniculata*	X	X
	Horsfieldia	*silvestris*	X	X
	Horsfieldia	*spicata*	X	
	Horsfieldia	*subtilis*	X	
	Myristica	aff. *subalulata*		X

FAMILY	GENUS	SPECIES	ALLUVIAL	HILL
	Myristica	archboldiana	X	
	Myristica	buchneriana	X	X
	Myristica	globosa		X
	Myristica	hollrungii	X	X
	Myristica	sp. A	X	
	Myristica	sp. B	X	
Myrsinaceae	*Rapanea*	aff. *leucantha*		X
Myrtaceae	*Decaspermum*	cf. *bracteatum*		X
	Decaspermum	fruticosum	X	
	Decaspermum	sp.		X
	Kania	eugenioides	X	X
	Lindsayomyrtus	brachyandrus		X
	Rhodamnia	pachyloba		X
	Rhodomyrtus	elegans	X	
	Rhodomyrtus	sp.	X	X
	Syzygium	aff. *furfuraceum*	X	X
	Syzygium	aff. *longipes*	X	X
	Syzygium	aff. *thornei*	X	X
	Syzygium	cf. *vernicosum*	X	
	Syzygium	acuminatissimum		X
	Syzygium	effusum	X	X
	Syzygium	fastigatum	X	X
	Syzygium	malaccense		X
	Syzygium	gonatanthum	X	
	Syzygium	laqueatum	X	X
	Syzygium	sp. nov.		X
	Syzygium	sp. A	X	X
	Syzygium	sp. B	X	
	Syzygium	sp. C	X	X
Ochnaceae	*Brackenridgea*	ridgia	X	
	Schuurmansia	sp.		X
Olacaceae	*Anacolosa*	papuana	X	
Oleaceae	*Chionanthus*	ramiflorus	X	

FAMILY	GENUS	SPECIES	ALLUVIAL	HILL
	Chionanthus	*rupicolus*		x
	Chionanthus	*sessiliflorus*	x	
	Chionanthus	sp.	x	
	Olea	aff. *javanica*	x	
Pandaceae	*Galearia*	*celebica*	x	x
Pandanaceae	*Pandanus*	sp.	x	x
Pittosporaceae	*Pittosporum*	aff. *ramiflorum*	x	
Podocarpaceae	*Decussocarpus*	*wallichianus*		x
Proteaceae	*Helicia*	*latifolia*		x
Rhamnaceae	*Alphitonia*	*incana*		x
Rhizophoraceae	*Gynotroches*	*axillaris*	x	x
Rosaceae	*Prunus*	*schlechteri*	x	x
Rubiaceae	*Canthium*	*cymigerum*		x
	Guettardella	aff. *erythrocarpa*	x	
	Guettardella	aff. *megacarpa*	x	
	Guettardella	sp.	x	
	Neonauclea	*gordoniana*	x	
	Psychotria	sp.	x	
	Rhadinopus	sp.	x	
	Tarenna	*pavetta*		x
	Timonius	aff. *timon*	x	x
	Timonius	sp.	x	
Rutaceae	*Acronychia*	aff. *acidula*	x	
	Euodia	aff. *bonwickii*	x	x
	Euodia	sp.	x	x
	Evodiella	sp.	x	
	Halfordia	*papuana*	x	x
	Tetractomia	*lauterbachiana*		x
Sabiaceae	aff. *Meliosma*	sp.	x	
Sapindaceae	*Alectryon*	*repandodentatus*	x	x
	Alectryon	sp.	x	
	Dictyoneura	*obtusa*	x	
	Guioa	sp.	x	

FAMILY	GENUS	SPECIES	ALLUVIAL	HILL
	Mischocarpus	aff. *largifolius*	X	
	Pometia	*pinnata*	X	X
Sapotaceae	*Palaquium*	sp.		X
	Pouteria	*anteridifelia*		X
	Pouteria	*firma*	X	
	Pouteria	*macropoda*		X
	Pouteria	*obovoides*		X
	Pouteria	*thrysoidea*	X	
	Pouteria	sp.	X	X
Sterculiaceae	*Sterculia*	*ampla*	X	X
	Sterculia	*macrophylla*	X	
	Sterculia	*schillinglawii*		X
Stilaginaceae	*Antidesma*	sp.	X	
Symplocaceae	*Symplocos*	*cochinchinensis*	X	X
	Symplocos	*cochinchinensis leptophylla*	X	
Theaceae	*Ternstroemia*	*cherryi*	X	
	Ternstroemia	sp.	X	
Tiliaceae	*Microcos*	*schlechteri*	X	
	Microcos	sp.	X	
	Microcos	*tetrasperma*	X	X
Verbenaceae	*Teijsmanniodendron*	*ahernianum*	X	X
	Teijsmanniodendron	*bogoriense*	X	X
Violaceae	*Rinorea*	*horneri*		X
Winteraceae	*Zygogynum*	sp.		X
Xanthophyllaceae	*Xanthophyllum*	*papuanum*	X	

List of Plants Collected by the Botanical Survey Team APPENDIX 3

List of plants collected during the 1996 Lakekamu Basin RAP survey. Numbers refer to collection voucher followed by annotations.

TAXA	EVIDENCE
ADIANTACEAE	
Pityrogramma calomelanos (L.) Link.	sight record from riverbank.
Syngramma schlechteri Brause	11487, alluvial forest. = *Craspedodictyum schlechteri* (Brause) Copel.
Taenitis hookeri (C. Chr.) Holttum	11315, alluvial forest.
ASPLENIACEAE	
Asplenium acrobryum C. Chr.	sight record from alluvial forest.
Asplenium amboinense Willd.	sight record from alluvial forest.
Asplenium bipinnatifidum Bak.	11603, alluvial forest.
Asplenium cuneatum Lam.	sight record of common epiphyte, uncertain relationship between this taxon and *A. affine* Swartz, alluvial forest.
Asplenium nidus L.	sight record of common fern throughout survey area.
Asplenium phyllitidus Don ssp. *malesicum* Holtt	sight record of common fern throughout survey area.
Asplenium regis Copel.	11461, coll. from alluvial forest but also occurring throughout the survey area.
Asplenium subemarginatum Ros.	11522, alluvial forest.
Asplenium tenerum Forst.	sight record, very common in alluvial forest.
ATHYRIACEAE	
Callipteris prolifera (Lam.) Bory	sight record of common fern in alluvial forest.
Callipteris spinulosa (Blume) J. Smith	11323, alluvial forest.
Diplazium cordifolium Blume	sight record of very common ground fern throughout the survey area.
BLECHNACEAE	
Blechnum orientale L.	10502, coll. from hill forest but occurring throughout the survey area.
Stenochlaena areolaris (Harr.) Copel.	sight record from alluvial forest.
Stenochlaena milnei Underw.	11523, robust and common climbing fern, mainly in alluvial forest.
Stenochlaena palustris (Burm. f.) Beddome	sight record from alluvial forest.

TAXA	EVIDENCE
CYATHEACEAE	
Cyathea contaminans (Wall.) Copel.	sight record of tree fern in seral situations, alluvial forest and hill forest.
Cyathea sp. A	11573, hill forest.
Cyathea sp. B	11649, hill forest, sp. with pinnate leaves, exindusiate.
Cystodium sorbifolium J. Sm.	sight record of common fern found throughout the survey area.
DAVALLIACEAE	
Davallia solida (Forst.) Sw.	sight record of common epiphyte in alluvial forest.
Humata heterophylla (Sm.) Desv.	sight record from riverbank.
Humata pusilla (Mett.) Carr	11567, hill forest.
Humata tenuis Copel.	11459, riverbank.
Leucostegia pallida (Mett.) Copel.	11356, alluvial forest.
DENNSTAEDTIACEAE	
Dennstaedtia sp.	11420, alluvial forest.
Hypolepis papuana F.M. Bailey	sight record, riverbank.
Sphenomeris retusa (Cav.) Maxon	11520, riverbank.
DRYOPTERIDACEAE	
Arachniodes sp.	11356, alluvial forest.
EQUISETACEAE	
Equisetum debile Roxb.	sight record from riverbank alluvium.
GLEICHENIACEAE	
Gleichenia linearis (Burm.) Underwood.	sight record of common fern in heliophytic regrowth situations (mainly riverbanks).
GRAMMITIDACEAE	
Loxogramme scolopendrioides (Gaud.) Morton	sight record from alluvial forest.
HYMENOPHYLLACEAE	
Hymenophyllum sp.	11310, alluvial forest.
Hymenophyllum sp.	11404, hill forest.
Pleuromanes pallidum (Bl.) Presl	sight record from hill forest.
Trichomanes aphlebioides Chr.	sight record from alluvial forest.
Trichomanes sp. (s.l.)	11317, alluvial forest.

TAXA	EVIDENCE
LINDSAEA GROUP	
Lindsaea bakeri (C. Chr.) C. Chr. var. *bakeri*	11489, alluvial forest.
Lindsaea lucida Bl. ssp. *brevipes* (Copel.) Kramer	11403, hill forest.
Lindsaea microstegia Copel.	11501, hill forest. neotenic form with short 1x pinnate lamina but soriferous.
Lindsaea microstegia Copel.	11560, hill forest. normal facies; intermixed simply pinnate and bipinnate fronds from common rhizome.
Lindsaea obtusa J. Smith	11410, hill forest. approaching *Lindsaea parallelogramma* v.A.v.R.
Lindsaea tenuifolia Bl.	sight record, fern in hill forest.
Tapeinidium longipinnulum (Cesati) C. Chr.	11460, alluvial forest.
LOMARIOPSIDACEAE	
Bolbitis heteroclita (Presl) Ching	11540, alluvial forest.
Bolbitis quoyana (Gaud.) Ching	sight record of common fern throughout the survey area.
Lomariopsis kingii (Copel.) Holttum	11521, common epiphytic fern mainly in alluvial forest.
LYCOPODIACEAE	
Lycopodium cernuum L.	sight record from heliophytic situations, hill forest.
Lycopodium nummularifolium Bl.	sight record from alluvial forest.
Lycopodium phlegmaria L.	11569, hill forest.
MARATTIACEAE	
Angiopteris evecta Hoffm.	sight record, common in alluvial forest esp. riverbanks.
Marattia sp.	11394, coll. from hill forest but also seen in alluvial forest.
OLEANDRACEAE	
Nephrolepis sp.	very common throughout the survey area, no colls.
Oleandra sibbaldii Presl	relatively common in the hill forest, cf. s.n. fragment coll. from ground.
OPHIOGLOSSACEAE	
Ophioglossum pendulum L.	one sighting in alluvial forest.
POLYPODIACEAE	
Belvisia mucronata (Fee) Copel. var. *mucronata*	11373, hill forest.

TAXA	EVIDENCE
Drynaria rigidula (Sw.) Bedd.	sight record of epiphyte in hill forest.
Drynaria sparsisora (Desv.) T. Moore	sight record from alluvial forest.
Goniophlebium percussum (Cauv.) Wagner & Grether	sight record of common fern in alluvial forest.
Microsorum punctatum (L.) Copel.	11582, alluvial forest.
Microsorum subgeminatum (Christ) Copeland	11366, alluvial forest.
Phymatosorus nigrescens (Bl) Picchi Sermolli	11610, alluvial forest = *Microsorum alternifolium* (Wildenow) Copeland.
Pyrrosia foveolata (Alston) Morton var. *lauterbachii* Hovenkamp	11364, riverbank.
Pyrrosia longifolia (Burm. f.) Morton	sightings in alluvial forest.
Pyrrosia princeps (Mett.) Morton	11365, riverbank.
PSILOTACEAE	
Psilotum complanatum Sw.	sight record from riverbank.
Psilotum nudum (L.) Gris.	sight record, hill forest.
PTERIDACEAE	
Pteris ligulata Gaud.	11519, riverbank.
Pteris tripartita Sw.	11517, riverbank.
SCHIZAEACEAE	
Lygodium salicifolium Presl.	11538, alluvial forest.
Schizaea dichotoma (L.) J. J. Sm.	sight record from alluvial forest.
SELAGINELLACEAE	
Selaginella hieronymiana v.A.v.R.	11535, alluvial forest.
Selaginella cf. *schlechteri* Hieron.	11532, alluvial forest.
Selaginella wariensis Hieron. vel. aff.	11488, alluvial forest.
TECTARIA GROUP	
Chlamydogramme hollrungi (Kuhn) Holttum	11483, hill forest.
Pleocnemia dahlii (Hieron.) Holttum	sight record from alluvial forest.
Tectaria beccariana (Cesati) C. Chr.	11570, hill forest and 11622 alluvial forest.
Tectaria pleiosora (Alderw) C. Chr.	sight record of common fern, pop. sterile, hill forest.
THELYPTERIDACEAE	
Metathelypteris polypodioides (Hook.) Holttum	11518, riverbank.
Sphaerostephanos unitus (L.) Holtt.	sight record from riverbank, common weed.

TAXA	EVIDENCE
VITTARIACEAE	
Antrophyum alatum Brack.	sight record from alluvial forest.
Vittaria angustifolia Bl.	11500, hill forest.
Vittaria elongata Sw. var. *angustifolia* Holtt.	11304, alluvial forest.
Vittaria scolopendrina (Bory) Thw.	11393, coll. from hill forest but occurring throughout the survey area.
GNETACEAE	
Gnetum gnemon L.	numerous sightings in alluvial forest.
Gnetum latifolium Bl.	numerous sightings in alluvial forest.
PODOCARPACEAE	
Podocarpus sp./spp.	occasional throughout the survey area; population(s) sterile.
AGAVACEAE	
Cordyline sp. (*fruticosa* or *terminalis*)	sight record from alluvial forest.
Dracaena angustifolia Roxb.	sight record of common sp. in alluvial forest.
ARACEAE	
Alocasia cf. *macrorrhizos* (L.) G. Don	sight record from alluvial forest.
Amydrium zippelianum (Schott) Nicolsen	sight record from alluvial forest.
Cyrtosperma (prob. *cuspidispathum* or *macrotum*)	sight record of sterile plants in alluvial forest.
Holochlamys beccarii Engl.	sight record, common sp. in alluvial forest.
Homalomena sp.	sight record, aromatic sterile herbs in riverbank forest.
Rhaphidophora sp. A	11431, alluvial forest.
Rhaphidophora sp. B	11525, alluvial forest.
Xenophya lancifolia (Engl) A. Hay	sight record, occasional in alluvial forest = *Alocasia lancifolia* Engl.
ARECACEAE/PALMAE	
Areca sp.	11598, alluvial forest.
Calamus aff. *brassii* Burret.	11594, alluvial forest.
Calamus humboldtianus Becc.	11587, alluvial forest.
Calamus schlechterianus Becc.	11588, alluvial forest.
Calamus sp.	11576, leaves lost in transit.

TAXA	EVIDENCE
Caryota rumphiana Bl. ex Mart. var. *papuana* Becc.	sight record of common sp. in alluvial forest.
Hydriastele sp.	11413, hill forest.
Korthalsia zippelii Bl.	sight record from alluvial forest.
Licuala sp.	11544, alluvial forest.
Linospadix albertisiana (Becc.) Burret	11391, hill forest.
Metroxylon sagu Rottb.	sight record of common sp. in alluvial forest.
Nengella sp.	11611, alluvial forest.
Ptychosperma sp.	11600, alluvial forest.
genus indet.	11478, hill forest.
CYPERACEAE	
Cyperus diffusus Vahl	11629, riverbank.
Cyperus cf. *cyperinus* (Retz.) Valck	11516, riverbank.
Hypoletrum nemorum (Vahl) Spreng. var. *nemorum*	11412, hill forest.
COSTACEAE	
Costus speciosus (Koen.) J. E. Sm.	sight record from riverbank.
DIOSCOREACEAE	
Dioscorea sp./spp.	sight record from alluvial forest, population(s) sterile.
FLAGELLARIACEAE	
Flagellaria indica L.	sight record from alluvial forest, population sterile.
LILIACEAE	
Crinum gracile E. Meyer in Presl	11653 (specimens lost in transit), sight determination based on narrow leaves (< 3.5 cm width), alluvial forest.
Dianella ensifolia (L.) DC.	sight record of common herb in hill forest understory.
MARANTACEAE	
Phacelophrynium sp.	11606, alluvial forest.
ORCHIDACEAE	
Agrostophyllum sp.	live coll. from alluvial forest.
Calanthe cf. *triplicata* (Willem) Ames	live coll. from alluvial forest.
Coelogyne beccarii Rchb. f.	11543, det. Howcraft 7/97, alluvial forest.
Dendrobium macrophyllum A. Rich.	11591, alluvial forest.

TAXA	EVIDENCE
Dendrobium cf. *biflorum* Sw.	11621, det. Howcroft, alluvial forest.
Dendrobium macrophyllum A. Rich	11591, det. WT, conf. Howcroft, alluvial forest.
Dendrobium smilliae F. Muell.	11372, det. WT, conf. Howcroft, hill forest.
Dipodium pandanum Bail.	sight record of climbing orchid along riverbank.
Grammatophyllum papuanum J. J. Sm.	11648, det. WT conf. Howcroft, hill forest.
Oberonia sp.	11452, riverbank.
Phreatia or *Thelasis* sp.	11369, riverbank.
Pseuderia aff. *frutex* or *foliosum*	11453, det. Howcroft, riverbank.
Robiquetia gracilistipes (Schltr.) J. J. Sm.	11592, alluvial forest.
PANDANACEAE	
Freycinetia angustissima Ridl.	11481, ex D. Bickford, alluvial forest.
Freycinetia klossii Ridl.	11542, alluvial forest.
Pandanus sp./spp.	sight records of sterile trees from alluvial forest.
POACEAE/GRAMINEAE	
Bambusa forbesii (Ridl.) Holttum	11534, alluvial forest.
Centotheca lappacea (L.) Desv.	sight record from alluvial forest.
Leptaspis urceolata (Roxb.) R. Br.	11322 and 11430, alluvial forest.
Pennisetum macrostachyum (Brogn.) Trin.	11586, riverbank.
Saccharum spontaneum L. (probably)	sight record from riverbed.
Thysanolaena maxima (Roxb.) O.K.	11429, riverbed.
genus indet.	11368, riverbank alluvium.
SMILACACEAE	
Smilax calophylla Wall. exDC.	11551, hill forest.
Smilax 'zeylanica group'	11564, hill forest.
ZINGIBERACEAE	
Curcuma australasica Hook. f.	11333, alluvial forest.
Pleuranthodium sp.	11482, hill forest.
Reidelia sp.	11342, alluvial forest.
Reidelia sp.	11433 and 11475, alluvial forest and hill forest.
Tapeinochilos sp.	sight record from alluvial forest, uncommon and sterile.
genus indet.	11527, alluvial forest.

TAXA	EVIDENCE
ACANTHACEAE	
Calophanoides angusta (Warb.) Bremek.	sight record, = *Calycacanthus magnusianum* K. Schum., common sp. in alluvial forest.
Hemigraphis reptans T. Anders. vel aff.	11536, herbs on riverbank boulders, alluvial forest.
ACTINIDIACEAE	
Saurauia aff. *stichophlebia* Diels	11607, alluvial forest.
AMARANTHACEAE	
Achyranthes sp. (*aspera* or *bidentata*)	sight record, alluvial forest.
ANACARDIACEAE	
Dracontomelon dao (Blanco) Merr. & Rolfe	11345, alluvial forest.
Rhus taitensis Guill.	sight record from riverbanks.
Semecarpus aruensis Engl.	11319, alluvial forest.
Semecarpus magnificus K. Schum.	11662, alluvial forest.
Semecarpus (closest to *?schlechteri* Laut)	11602, alluvial forest.
ANNONACEAE	
Artabotrys suaveolens (Bl.) Bl.	11530, (petals to > 2 cm), streambank.
Cananga odorata (Lamk) Hook. f. & Thoms.	sight record from alluvial forest, pop. flowering.
Cyathocalyx papuanus Diels	11615, alluvial forest.
Haplostichanthus longirostris (Scheffer) van Heusden	11339, alluvial forest.
Mitrella kentii (Bl.) Miq. or aff.	11449, riverbank.
Polyalthia aff. '*michaelii* group'	11416, alluvial forest.
Popowia pisocarpa (Bl.) Endl.	11307 and 11640, alluvial and hill forest.
Pseuduvaria grandifolia (Warb.) Sinclair or aff.	11616, alluvial forest.
Pseuduvaria aff. *?versteegii* (Diels) Merr.	11617, alluvial forest.
Pseuduvaria or *Popowia* sp.	11628, alluvial forest.
Pseuduvaria sp.	11633, alluvial forest.
Uvaria cf. *rosenbergiana* Scheffer	11450, riverbank.
Xylopia aff. *caudata* Hook. f. & Thoms.	11636, hill forest.
Xylopia malayana Hook. f. & Thoms.	11419, alluvial forest.
APOCYNACEAE	
Alstonia sp.	sight record of uncommon sp. in alluvial forest, sterile plants.

TAXA	EVIDENCE
Alyxia cf. *maluensis* Markgr.	11480, hill forest.
Alyxia markgrafii Tsiang	11660, specimens lost in transit, probably equivalent to 11740, alluvial forest.
Neisosperma ficifolium (Sp. Moore) Fosb. & Sach.	11421, specimens lost in transit but was apparently equivalent to 11237, alluvial forest.
Parsonia aff. *?brassii* Markgr.	11639, hill forest.
Parsonia oligantha (K. Schum.) D. J. Middleton	11326, from riverbank, = *Delphyodon oliganthus* K. Schum.
Wrightia laevis Hook. f. ssp. *novoguineensis* P. t. Ngan	11474, hill forest.

ARALIACEAE

Osmoxylon novoguineense (Scheff.) Becc.	sight record from alluvial forest.
Schefflera (2 spp.)	sightings from hill forest.

ARISTOLOCHIACEAE

Aristolochia or *Pararistolochia* sp.	sight record of sterile populations throughout the survey area.

ASCLEPIADACEAE

Cyrtolepis ?sp. nov.	11451, does not key out on Forster, riverbank.
Dischidia sp.	11638, hill forest.
Hoya sp.	11352, alluvial forest.

ASTERACEAE/COMPOSITAE

Bidens pilosa L.	sight record of weed along riverbank.
Blumea arfakiana Martelli	sight record of riverbank weed.
Crassocephalum crepidioides (Benth.) S. Moore	sight record from riverbank.

BARRINGTONIACEAE

Barringtonia sp.(*calyptrocalyx* group)	sight record from alluvial forest.
Barringtonia sp.(*leptocaul* group)	sight record from alluvial forest.
Planchonia papuana Laut.	sight record of common sp. from alluvial forest and riverbanks.

BEGONIACEAE

Begonia aff. 'media group'	11599, alluvial forest.

BIGNONIACEAE

Tecomanthe dendrophila (Bl.) K. Schum. & Laut.	sight record from alluvial forest.

TAXA	EVIDENCE
BIXACEAE	
Bixa orellana L.	sight record from riverbank.
BUDDLEIACEAE	
Buddleia asiatica Lour.	sight record of riverbank weed.
BURSERACEAE	
Canarium asperum Benth.	11537, alluvial forest.
Canarium indicum L.	sight record from alluvial forest.
Canarium vitense A. Gray	11330, alluvial forest.
CASUARINACEAE	
Gymnostoma sp.	sight record from alluvial forest and riverbanks.
CLUSIACEAE/GUTTIFERAE	
Calophyllum euryphyllum Laut.	11374 collection from ground, hill forest.
Calophyllum euryphyllum Laut.	11386, det. by P. F. Stevens (Apr. 1997), hill forest.
Calophyllum 'goniocarpum complex'	11407, det. by P. F. Stevens (Apr. 1997), hill forest.
Calophyllum 'goniocarpum complex'	11497, det. by P. F. Stevens (Apr. 1997), hill forest.
Calophyllum papuanum Laut.	11554, det. by P. F. Stevens (Apr. 1997), hill forest.
Calophyllum soulattri Burm. f.	11485, (sterile voucher), alluvial forest.
Garcinia maluensis Laut.	11463, hill forest.
Garcinia sp. A	11507, hill forest.
Garcinia sp. B	11379, 11492, seen by P. F. Stevens without det., hill forest.
Mammea cordata P. F. Stevens	11624, alluvial forest.
COMBRETACEAE	
Quisqualis indica L.	11425, specimen lost in transit, alluvial forest.
Terminalia impediens Coode	sight record of common alluvial forest sp.
Terminalia rubiginosa K. Schum.	11441, alluvial forest.
CONNARACEAE	
Rourea radlkoferiana K. Schum.	11644, hill forest.
CONVOLVULACEAE	
Merremia sp. (probably *peltata*)	sight record from riverbank.
CUNONIACEAE	
Caldcluvia nymani (K. Schum.) Hoogl.	11325, alluvial forest.
DATISCACEAE	
Octomeles sumatrana Miq.	11328, alluvial forest.

TAXA	EVIDENCE
DICHAPETALACEAE	
Dichapetalum sp.	11545, alluvial forest.
DILLENIACEAE	
Dillenia sp.	sight record of sterile trees from alluvial forest.
Tetracera nordtiana F.v.M.	sight record, common riverbank climber.
ELAEOCARPACEAE	
Aceratium cf. *oppositifolium* DC.	11558, hill forest.
Elaeocarpus blepharoceras Schltr.	11494 (flowering) and 11664 (fruiting), hill forest.
Elaeocarpus dolichostylus Schltr. ssp. *dolichostylus*	11346, alluvial forest.
Elaeocarpus ledermannii Schltr.	11571, hill forest.
Elaeocarpus sepikianus Schltr.	11546, hill forest.
Sloanea nymani K. Schum.	11613, alluvial forest.
Sloanea paradisearum F. Muell.	11595, alluvial forest.
ERICACEAE	
Dimorphanthera sp.	11498, det by P. F. Stevens as sp. nov. or var. nov. aff. *breviflos*; tentative manuscript name *D. biglandulosa* hill forest.
Rhododendron sp.	11503, hill forest.
EUPHORBIACEAE	
Acalypha hellwigii Warburg	sight record from alluvial forest, same sp as 11197.
Antidesma contractum J. J. Sm.	11458, hill forest.
Antidesma sphaerocarpum Muell. Arg.	11428 and 11510, alluvial forest.
Antidesma sphaerocarpum Muell. Arg.	11472, hill forest.
Aporusa petiolaris Airy Shaw	11572, hill forest.
Baccaurea philippinensis (Merr.) Merr.	11414 and 11439, ditto, male plant ex A. Mack 11395, alluvial forest.
Breynia cernua (Poir.) Muell.-Arg.	sight record from alluvial forest.
Bridelia glauca Bl.	11355, alluvial forest.
Claoxylon cf. *microcarpum* Airy Shaw, *ledermannii-microcarpum* group'	11313, alluvial forest.
Endospermum sp.	sight records from alluvial forest, population sterile.
Exocoeria indica (Wild.) Muell. Arg.	sight record from alluvial forest.
Glochidion novo-guinense K. Schum.	11363, riverbank.
Macaranga aleuritoides F. Muell.	11585, riverbank.

TAXA	EVIDENCE
Macaranga bifoveata J. J. Sm.	11514, riverbank.
Macaranga fallacina Pax & Hoffm.	11609, alluvial forest.
Macaranga polyadenia Pax & Hoffm.	11347, alluvial forest.
Macaranga punctata K. Schum.	11593, alluvial forest.
Macaranga tanarius (L.) Muell. Arg.	sight record from alluvial forest.
Macaranga aff. *warburgiana* Pax & Hoffm.	11443, alluvial forest.
Octospermum pleiogynum (P. & H.) Airy Shaw	11635, specimen lost in transit, alluvial forest.
Phyllanthus rheophilus Airy Shaw	11455, det. by K. Demas, rheophyte on riverbank.
Phyllanthus sp.	11340, alluvial forest.
Pimelodendron amboinicum Hassk.	11427, alluvial forest.

FABACEAE/LEGUMINOSAE

Caesalpinioideae

Bauhinia ampla Span. var. *schlechteri* (Harms) K. Larsen & Sunarno	11506, base camp, alluvial forest.
Maniltoa fortuna-tironis Verdc.	11351, keys to species on Verdcourt., riverbank forest.
Maniltoa fortuna-tironis Verdc.	11541, keys to species on Verdcourt., alluvial forest.
Maniltoa fortuna-tironis Verdc	11643, keys to species on Hou et al., hill forest.
Maniltoa plurijuga Merr. & Perry	sight record, numerously jugate sp., common in alluvial forest.
Phaseolus sp. (sensu lato)	11531, alluvial forest.

Mimosoideae

Archidendron clypearia (Jack) Nielsen	11385, =*Abarema clypearia* (Jack) Kosterm., hill forest.
Archidendron sp.	11348 and 11650, alluvial forest.
Entada sp. (*phaseoloides* or *pursaetha*)	sight record from throughout the survey area.

Papilionoideae

Centrosema pubescens Benth.	sight record, weedy vine in streambed.
Crotalaria anagyroides H. B. K.	11580, weedy shrub from South America on riverbanks.
Derris elegans Grah. ex Benth. var. *elegans*	11359, riverbank.
Pterocarpus indicus Willd.	sight record, common sp. in alluvial forest esp. riverbanks.
Pueraria lobata (Willd.) Ohwi	sight record of weedy vine along riverbanks.

FAGACEAE

Lithocarpus celebicus (Miq.) Rehd.	sight record from alluvial forest.

TAXA	EVIDENCE
Lithocarpus cf. *schlechteri* Markgr.	11468, hill forest.

FLACOURTIACEAE

Casearia clutiaefolia Bl.	sight record from alluvial forest.
Homalium foetidum (Roxb.) Benth	11577, alluvial forest.
Pangium edule Reinw.	sight record from alluvial forest.

GESNERIACEAE

Aeschynanthus sp.	11353, testa papillose, seeds ellipsoid, plumose at one pole, caudate at the opposite pole, alluvial forest.
Cyrtandra aff. *bracteata* Warburg	sight record from alluvial forest.
Cyrtandra aff. *janowskyi*	sight record from hill forest.
Cyrtandra sp. A	11314, alluvial forest.
Cyrtandra sp. B	11332, alluvial forest.
Cyrtandra sp. C	11579, alluvial forest.

GOODENIACEAE

Scaevola oppositifolia R. Br.	sight record from alluvial forest.

GROSSULARIACEAE

Polyosma 'cestroides-induta group'	11418, alluvial forest.

HERNANDIACEAE

Hernandia ovigera L.	sight record, occasional sp. in alluvial forest.

HYDRANGEACEAE

Dichroa sylvatica (Reinw. ex Bl.) Merr.	11457 and 11605, alluvial forest.

ICACINACEAE

Gomphandra papuana (Becc.) Sleum.	11311, alluvial forest.
Rhyticaryum longifolium K. Schum. & Laut.	11657, coll. from riverbank.
Rhyticaryum cf. *novoguineensis* (Warb.) Sleumer	11462, riverbank in hill forest.

LAURACEAE

Actinodaphne nitida Teschn.	11476, hill forest.
Actinodaphne (*nitida* group)	sight record, sterile population, alluvial forest.
Beilschmiedia acutifolia Teschn.	11378, hill forest.
Beilschmiedia cf. *schoddei* Kosterm. 'acutifolia group'	11380, hill forest.
Cryptocarya depressa Warb.	11529, alluvial forest.
Cryptocarya cf. *novoguineensis* Teschn.	11324, alluvial forest.

TAXA	EVIDENCE
Cryptocarya cf. '*viridiflora* group'	11387, hill forest.
Cryptocarya sp. A	11381, hill forest.
Cryptocarya sp. B	11400, indumentum of coarse appressed hairs; needs to be checked, hill forest.
Litsea firma (Bl.) Hook. f.	11553, hill forest.
Litsea aff. '*grandiflora* group'	11596, allied to the large-leaved forms, but cannot match the setiform indumentum, ? sp. nov., alluvial forest.
LEEACEAE	
Leea coryphantha Laut.	11422, alluvial forest.
Leea indica (Burm. f.) Merr.	11539, riverbank forest.
Leea zippeliana Miq.	11597, alluvial forest.
LINACEAE	
Hugonia jenkinsii F.v.M.	11327, alluvial forest.
LOGANIACEAE	
Fagraea ceilanica Thunb.	11469, hill forest.
Fagraea elliptica Roxb.	sight record from riverbanks.
Fagraea racemosa Jack ex Wall.	sight record of common sp. from alluvial forest.
Fagraea woodiana F. v. Muell.	11467, hill forest.
Neuburgia corynocarpa (A. Gray) Leenh.	11367, riverbank.
Neuburgia kochii (Val.) Leenh.	11358, alluvial forest.
LORANTHACEAE	
Decaisnina hollrungii (K. Schum.) Barlow	11392 and 11440, colls. from alluvial forest but occurring throughout the survey area.
LYTHRACEAE	
Lagerstroemia piriformis Koehne	11513, coll. from riverbank forest.
MALVACEAE	
Abelmoschus moschatus Medic.	sight record of trailside weed.
Abroma augusta (L.) Willd.	sight record from riverbanks.
Hibiscus archboldianus Bross.	11432 and 11590, alluvial forest.
Hibiscus sp.	11590, alluvial forest.
Thespesia populnea (L.) Sol. ex Corr.	sight record from alluvial forest and riverbanks.
MELASTOMATACEAE	
Astronidium sp. A	11477, hill forest.

TAXA	EVIDENCE
Astronidium sp. B	11508, riverbank.
Dissochaeta angiensis Ohwi	11438, alluvial forest.
Medinilla aff. *compacta* Bakh. f.	11526, alluvial forest.
Medinilla cf. *hollrungiana* Mansf.	11504, hill forest.
Medinilla rubrifructus Ohwi	11505, hill forest.
Medinilla cf. *versteegii* Mansf.	11371, riverbank.
Medinilla aff. '*versteegii* group'	11409 and 11552, hill forest.
Medinilla ?sp. nov.	11398, hill forest.
Memecylon schraderbergense Mansf.	11401B, hill forest.
Memecylon torricellense Mansf.	11528, alluvial forest.
Otanthera cyanoides Triana	11511, riverbank.
Poikilogyne aff. *hirta* sensu Streimann & Stevens	11396 and 11561, hill forest.
Poikilogyne aff. *villosa* Maxw.	11343 alluvial forest and 11464 on riverbank.
Pternandra rostrata (Cogn.) Nayar	11377, hill forest.

MELIACEAE

Aglaia agglomerata Merr. & Perry	11484, hill forest.
Aglaia subminutiflora C. de Candolle	11411, hill forest.
Aphanamixis polystachya (Wall.) R.N. Parker	11336, alluvial forest.
Dysoxylum alliaceum (Bl.) Bl.	11471, hill forest.
Dysoxylum alliaceum (Bl.) Bl vel aff. '*alliaceum* group'	11499 and 11555, hill forest.
Dysoxylum inopinatum (Harms) Mabb.	11495, hill forest.
Dysoxylum aff. '*papuanum* group'	11375, hill forest.
Dysoxylum phaeotrichum Harms	11646, hill forest.
Sandoricum koetjape (Burm, f.) Merr.	probable, specimen 11637 lost in transit, hill forest.

MENISPERMACEAE

Arcangelisia tympanopoda (Laut. & K. Schum.) Diels	11583, alluvial forest.
Chlaenandra ovata Miq.	endocarp collection from alluvial forest.
Hyserpa polyandra Becc. var. *polyandra*	11305, alluvial forest.
Parabaena sp.	11426, specimen lost in transit, alluvial forest.
Stephania japonica (Thunb. ex Murr.) Miers	sight record from alluvial forest.

TAXA	EVIDENCE
MONIMIACEAE	
Kibara archboldiana A. C. Smith	11533, alluvial forest.
Levieria acuminata (F. v. M.) Perkins	11382, 11548, and 11566, hill forest.
Palmeria gracilis Perkins	s.n. from sterile coll., but indumentum is diagnostic, hill forest.
Steganthera hospitans (Becc.) Kanehira & Hatusima	sight record from alluvial forest.
MORACEAE	
Artocarpus sp.	sight record, population sterile, alluvial forest.
Ficus arbuscula Laut. & K. Schum.	sight record from riverbanks.
Ficus botryocarpa Miq.	sight record from alluvial forest.
Ficus hesperidiformis King	fruits coll. by A. L. Mack, det. by G. Weiblen, alluvial forest.
Ficus ochrochlora Ridley	11362, riverbank.
Ficus odoardi King	11434, alluvial forest.
Ficus pachystemon Warb.	11338, alluvial forest.
Ficus pungens Reinw. ex Blume	sight record from alluvial forest.
Ficus subulata Bl.	11357, alluvial forest.
Ficus xylosycia Diels var. *cylindrocarpa* (Diels) Corner	11337, alluvial forest.
MYRISTICACEAE	
Gymnacranthera paniculata (DC.) Warb. var. *zippeliana* (Miq.) J. Sinclair	11642, hill forest.
Horsfieldia subtilis (Miq.) Warb. var. *subtilis*	11331, alluvial forest.
Myristica buchneriana Warb.	11445, alluvial forest.
Myristica lancifolia Poir.	11384 and 11470, hill forest.
Myristica subalulata Miq.	sight record of a common sp. mainly from alluvial forest.
MYRSINACEAE	
Ardisia sect. *Acrardisia*, poss. sp. nov.	11312, alluvial forest.
Grenacheria buxifolia Mez	11493, hill forest.
Rapanea acrosticta Mez, poss. sp. nov.	11612, alluvial forest.
Rapanea lamii Sleumer	11473, hill forest.
MYRTACEAE	
Acmena acuminatissima (Bl.) Merr. & Perry	11496, hill forest.

TAXA	EVIDENCE
Decaspermum bracteatum (Roxb.) A. J. Scott	sight record, (but rachis subglabrous), alluvial forest.
Kania eugenioides Schltr.	11491, hill forest.
Metrosideros cf. *ramiflora* Laut.	11383, hill forest.
Rhodomyrtus trineura (F. Muell.) F. Muell. ex Benth var. *novoguineensis* (Diels) A. J. Scott	11390, (note: as *Rhodomyrtus novoguineensis* Diels on Guymer's revision), hill forest.
Syzygium hylophilum (Laut. & K. Schum.) Merr. & Perry	11341, alluvial forest.
Syzygium longipes (Diels) Merr. & Perry	11479, from hill and 11618 alluvial forest.
Syzygium aff. '*nutans-insulare* group'	11627, alluvial forest.
Syzygium rosaceum Diels or aff.	11401, hill forest.
Syzygium aff. '*schumannianum* group'	11550, (related to taxa with linear-clavate calyx tube), hill forest.
Syzygium thornei Hartley & Perry	11563, keys out on Hartley and Perry, hill forest.
Syzygium trivene (Ridley) Merr. & Perry	11448, riverbank.
Syzygium sp. nov. aff. *cladopterum-madangense* group	11601, very unusual; large cordate-based leaves with inflo. in large mounds at ground level, alluvial forest.
Syzygium sp.	11399, hill forest.
NYCTAGINACEAE	
Pisonia longirostris Teys. & Binn.	11424, alluvial forest.
OCHNACEAE	
Schuurmansia henningsii K. Schum.	sight record of sp. common throughout the survey area.
PASSIFLORACEAE	
Adenia cf. *heterophylla* (Bl.) Koord.	11661, specimen lost in transit, coll. from riverbank.
Passiflora foetida L.	sighting on riverbank.
PIPERACEAE	
Piper mestonii F.M. Bailey	sight record from riverbank.
Piper novo-guineense Warb.	11515, riverbank.
Piper subcanirameum C. DC.	11632, alluvial forest.
PITTOSPORACEAE	
Pittosporum ramiflorum (Zoll. & Mor.) Zoll.	11447, alluvial forest.
Pittosporum sinuatum Bl.	sight record from alluvial forest.
POLYGALACEAE	
Eriandra fragrans Royen & Steenis	11335 and 11406, alluvial forest and hill forest.

TAXA	EVIDENCE
Polygala paniculata L.	sight record of weed in disturbed forest margin.
Securidaca ecristata Kassau	11565 hill forest and 11578 alluvial forest.
PROTEACEAE	
Helicia sp.	sight record from alluvial forest.
RHAMNACEAE	
Alphitonia macrocarpa Mansf.	11334, alluvial forest.
Emmenosperma alphitonioides F. Muell.	sight record from hill forest.
Gouania leptostachya DC.	11575, alluvial forest.
Ventilago papuana Merr. & Perry	11574, riverbank.
Zizyphus djamuensis Laut.	11444, riverbank.
RHIZOPHORACEAE	
Carallia brachiata (Lour.) Merr.	11370, riverbank.
Gynotroches axillaris Bl.	sight record of common sp. in alluvial forest.
ROSACEAE	
Prunus arborea (Bl.) Kalk.	11549, hill forest.
Prunus dolichobotrys (K. Schum. & Laut.)	11320 and 11360, alluvial forest and riverbanks.
RUBIACEAE	
Canthium cymigerum (Val.) B. L. Burtt.	11435, alluvial forest.
Gardenia sp. (*lamingtonia* group)	sight record from alluvial forest.
Hedyotis aff. *auricularia* L.	11321, alluvial forest.
Lasianthus sp.	11397, hill forest.
Lucinaea sp.	11454, riverbank.
Morinda umbellata L.	sight record from riverbank.
Mussaenda bevani F.v.M.	11344, alluvial forest.
Mussaenda ferruginea K. Schum.	11466, alluvial forest.
Mussaenda scratchleyi Wernh.	sight record of common riverbank sp.
Mycetia javanica (Bl.) Reinw. ex Korth.	11316, alluvial forest.
Psychotria leptothyrsa Miq.	11405, hill forest.
Psychotria '*multicostata* group'	11604, alluvial forest.
Psychotria (vining sp.)	11562, 11645, unrevised group-cannot match, hill forest.
cf. *Psychotria*	11318, filiform peduncle, laxly cymose inflorescence, alluvial forest.
Randia decora Val.	11354, alluvial forest.

TAXA	EVIDENCE
Randia schumanniana Merrill & Perry	sight record from alluvial forest.
Randia?	11630, does not conform to any species, alluvial forest.
Tarenna buruensis (Miq.) Valeton	11631, alluvial forest.
Timonius avensis Val. or aff.	11436, alluvial forest.
Timonius aff. *grandifolia* Valeton	coll. from alluvial forest.
Uncaria callophylla Bl. ex. Korth.	11389, hill forest.
Urophyllum 'glaucescens-rostratum group'	11309 and 11417, closest to ?*glaucescens* Val. alluvial forest.
Urophyllum sp.	11417, alluvial forest.
cf. *Urophyllum*	11318, alluvial forest.
RUTACEAE	
Euodia cf. *hortensis* sJ. R. & G. Forst. or aff.	11557, hill forest.
Lunasia amara Blanco	11456, alluvial forest.
Tetractomia tetrandrum (Roxb.) Merr.	11402, hill forest.
SABIACEAE	
Sabia pauciflora Blume	sight record from riverbank.
SANTALACEAE	
Dendromyza cf. *ledermannii* (Pilger) Stauffer	11556, hill forest.
SAPINDACEAE	
Cupaniopsis cf. *macropetala* Radlk.	11656, riverbank.
Dictyoneura obtusa Bl.	sight record from alluvial forest, populations sterile.
Harpullia ramiflora Radlk.	11361, riverbank.
Lepisanthes senegalensis (Poir.) Leenh.	11524, alluvial forest.
Pometia pinnata Forst.	sight record, very common in alluvial forest.
SAPOTACEAE	
Burckella sp.	11388, hill forest.
Planchonella sp.	11423, alluvial forest.
cf. *Pouteria* sp.	11329, alluvial forest.
SAXIFRAGACEAE	
Dichroa sylvatica (Reinw. ex Bl.) Merr.	11457 and 11605, alluvial forest.
SOLANACEAE	
Solanum oliverianum Laut. & K. Schum. (*Lycianthes*).	11641, hill forest.

TAXA	EVIDENCE
STERCULIACEAE	
Melochia sp. (*odorata* or *umbellata*)	sight record from hill forest.
Pterocymbium beccarii K. Schum.	11614, alluvial forest.
Sterculia macrophylla Vent.	sight record from alluvial forest.
Sterculia cf. *shillinglawii* F. Muell.	coll. from alluvial forest.
Sterculia schummaniana (Laut.) Mildbr.	11465, riverbank.
THEACEAE	
Eurya roemeri Laut.	11350, also apparently as *E. tigang* Schum. & Laut. var. *roemeri* (Laut.) Barker, alluvial forest.
Ternstroemia britteniana F. v. M. or aff.	11408, hill forest.
Ternstroemia cherryi (F.M. Bail.) Merr.	11486, alluvial forest.
TILIACEAE	
Trichospermum sp. nov.	11589, does not key out on Kostermans, coll. from riverbanks.
Microcos tetrasperma Merr. & Perry	11415, alluvial forest.
ULMACEAE	
Gironniera celtidifolia Gaudich.	11308, alluvial forest.
Trema cannabina Lour.	11559, hill forest.
URTICACEAE	
Cypholophus sect. *Foliosae*, aff. *vestitus* (Bl.) Miq.	11626, keys to this result on Winkler but cannot confirm against authentically annotated material, alluvial forest.
Elatostema integrifolia Wedd.	sight record, = *Elatostema sesquifolium* (Reinw. ex Bl.) Hassk., alluvial forest.
Elatostema aff. *macrophyllum* Brongn.	sight record equivalent to 11229 Crater Mt., streambanks.
Elatostema weinlandii K. Schum.	11581, alluvial forest.
Leucosyke capitellata (Poir) Chew	sight record from riverbank.
Maotia cf. *ambigua* Wedd.	11584, riverbank.
Nothocnide melastomatifolia (K. Schum.) Chew	11437, alluvial forest.
Procris frutescens (Winkler) Schroter	11512, alluvial forest.
VERBENACEAE	
Callicarpa cf. *longifolia* Lam.	11608, but corolla is pink (not white) and glabrous, alluvial forest.

TAXA	EVIDENCE
Slerodendrum brassi Beer & H. J. Lam	11623, alluvial forest.
Clerodendrum tracyanum (F. Muell.) Benth.	11651, not *C. buruanum*, calyx adaxially with glandular hairs, riverbanks.
Geunsia pentandra (Roxb.) Merr.	sight record, = *G. farinosa* Bl. (Backer & Bakhuizen 1965) or as *Callicarpa*, alluvial forest.
Petraeovitex multiflora (J. E. Smith) Merr.	11619, alluvial forest.
Premna serratifolia L.	11446, alluvial forest.
Stachytarpheta sp.	sight record of riverbank weed.
Teijsmanniodendron ahernianum (Merr.) Bakhuizen	11376, hill forest.
VIOLACEAE	
Rinorea horneri (Korth.) O. Ktze.	11625, alluvial forest.
VITACEAE	
Ampelocissus muelleriana Planch. or aff.	11490, alluvial forest.
Cayratia or *Nothocissus* sp.	11620, alluvial forest.
Tetrastigma papillosum (Bl.) Planch.	11509, riverbank.
WINTERACEAE	
Zygogynum aff. *glaucum* (A. C. Smith) Vink	11547 and 11568, hill forest, apparently the same sp. also in alluvial forest.
FAMILY INDET.	
vine sp.	11306, 11442, alluvial forest.

Significant Botanical Records

Annonaceae

Xylopia aff. *caudata* Hook. f. &Thoms.; coll. 11636. Unusual characteristics are the small linear-elliptic leaves with silvery and sericeous indumentum. Not matching any Papuasian collection at Lae Herbarium. Closest to *Xylopia caudata* on Sinclair's (1955) account of Malayan Annonaceae. Possibly a new species, but in an unrevised genus in an unrevised family notorious for taxonomic confusion, who can be sure?

Apocynaceae

Wrightia laevis Hook. f. ssp. *novoguineensis* P.t. Ngan; coll. 11474. Although *Wrightia laevis* is widely distributed from China to Australia, the endemic ssp. *novoguineensis* is an apparently rare taxon, judging from the scarcity of herbarium collections (Ngan 1965). Lae Herbarium has only two sheets of this sub-species. The compound coronal segments, much exceeding the stamens, quickly separate ssp. *novoguineensis* from the more common ssp. *milgar.* In spite of its supposed rarity, the plant is abundant in the hill forest near the Lakekamu research house. Populations were flowering profusely during the RAP expedition.

Asclepiadaceae

Cryptolepis sp. nov.; coll. 11451. The 7 Malesian species have been thoroughly assessed in a series of papers by Forster (1990, 1991, 1993). Many of the taxa are seldom collected. The RAP number keys to *C. nymanii*, a species endemic to Morobe and Central Provinces, but is not that plant. Although the corolla is internally epapillate, the inflorescence is sparingly branched and densely bracteate. Moreover the leaf venation density is comparable to *C. multinervosa.* A very annoying specimen because the facies is individually composed of distinctive and recognizable elements, but which collectively cannot be fitted to any one species. Submitted to P. Foster of Queensland Herbarium for comment and annotation.

Caesalpiniaceae (Leguminosae)

Bauhinia ampla Span. var. *schlechteri* (Harms) K. Larsen & Sunarno; coll. 11506. = *Gigasiphon schlechteri* (Harms) de Wit. *Bauhinia* was recently revised in the Flora Malesiana (Hou et al. 1996) so taxonomic information on this group is very current. Collection 11506 from a large tree in camp, represents a species which is rare throughout its range. Two varieties are recognized. Variety *schlechteri* is endemic to New Guinea and was known from only two localities: Albatros Bivak in Irian Jaya, and Madang (ibid). Lae Herbarium has very few sheets from which the key result could be confirmed, but the entire leaves of 11506 are very unusual for the subgenus and taxonomic assignment was thus a relatively simple matter. The gathering is a first record for Papua and one of the few specimens ever obtained of this conspicuous species. The flowers are so large and striking that the scarcity of herbarium material probably derives from actual biological rarity rather than simple undercollecting.

Combretaceae

Terminalia rubiginosa K. Schum.; coll. 11441. Known to occur throughout Papua, except for Gulf Province, where it was nonetheless presumed to occur (Coode 1978). Now confirmed for the province.

Elaeocarpaceae

Elaeocarpus blepharoceras Schltr.; 11494 (flowering), 11664 (fruiting), xylarium and carpological specimens also. This is a very distinctive *Elaeocarpus*, and our collections conform in detail to the description provided by Coode (1981), even to the points made in regards to the remarkable features of the bark (as seen on the xylarium specimen accompanying 11664). Coode's comments on the marked resemblance to *Litsea* is also uncannily evocative of our own confusion in the field with sterile material of this taxon.

Elaeocarpus blepharoceras was first discovered in the Hunstein district, with most of the subsequent collections coming from Mt. Kaindi near Wau. The other known locality is Mt. Tafa in Central Province (ibid). The species is found in montane forest between 1800 - 2400 m elevation and has not been recorded below 1000 m. Until now, it was unknown from the Papuan lowlands and was almost exclusively a Mamose region montane species. The RAP collections from 175 m elevation are very anomalous in light of the ecology of this species.

Sloanea paradisearum F. Muell.; coll. 11595. Rare in Mamose region, *Sloanea paradisearum* had otherwise been collected from every Papuan district except Gulf Province (Coode 1981). Now recorded for Gulf.

Ericaceae

Dimorphanthera sp. nov.; coll. 11498. The RAP collection has been determined by Ericaceae specialist P. F. Stevens as species nov. or a variety nov. aff. *breviflos*. The material is being prepared for formal description and publication under the tentative manuscript name *D. biglandulosa* (Stevens, personal communication). *Dimorphanthera* is a montane genus of mossy forests; the present collection from below 150 m elevation is very atypical.

Euphorbiaceae

Baccaurea philippinensis (Merr.) Merr.; coll. 11395 ex Andy Mack (male plant), 11414 and 11439 (female plants). An indumentum of minute stellate hairs makes for an easy assignment on Airy Shaw (1980). *Baccaurea philippinensis* is a species of Western Malesia, and its presence in Papua New Guinea is based upon just two specimens, both from Western Province (ibid). One of the cited sheets was fortuitously found at Lae Herbarium in an indet. folder and is a good match for the present material. Airy Shaw commented on the close relationship between this species and *B. bracteata*, and the similarity is very apparent on LAE exsiccatae, except for the absence of black punctulations in *B. philippinensis* as noted by Shaw. Although the species has been rarely collected in PNG, it is reportedly common in the Lake Murray area (Pullen 7469). There is a large populations at Lakekamu (where the plant was seen numerous times), with definite ecological preference for riverbanks and alluvial flats.

Bridelia glauca Bl.; coll. 11355. A species with an aggregate range covering nearly the entire breadth of Malesia, but rare east of the Moluccas. The taxon is recorded from Papuasia by a single specimen (NGF 10976) from New Britain and has not been reported from the New Guinea mainland (Airy Shaw 1980). *Bridelia glauca* is distinguished from the ubiquitous *B. penangiana* by the glaucescent leaves, a character exhibited on collection 11355. Another point of distinction is the noticeably pedicelled fruits of the former species (sessile on *B. penangiana*). Taxonomic confirmation of the Lakekamu population can be accomplished by acquisition of fruiting material.

Macaranga aff. *warburgiana* Pax & Hoffm.; coll. 11593. An intriguing collection in Whitmore's 'dioica group' of species with long scapose infructescences. The leaves on 11593 are truncate, with basiscopic nervation from the proximal laterals and elongate glands at the petiole insertion. Apparently closest to *M. warburgiana*. There is a conspecific sheet at LAE with male inflorescences (NGF 46370) from Central Province, annotated by Airy Shaw as cf. *warburgiana*. However our collection has fruits, and the excrescences are completely unlike *warburgiana* s. str. Perhaps an infraspecific novelty or a new Papuan satellite species?

Melastomataceae

Medinilla sp. (?sp. nov.); coll. 11398. *Medinilla* is a genus similar to *Psychotria* in its taxonomic pattern of diversification into numerous localized endemics. Mansfeld revised the Papuasian taxa in 1925, and Merrill and Perry (1943) subsequently produced a key incorporating their own taxa into the Mansfeld conspectus. Numerous contemporary collections however, cannot be accommodated by either review, and clearly there are many undescribed species. The Lakekamu collection is a distinctive *Medinilla* with large linear-elliptic leaves and diffuse cauline panicles. Branches are quadrangular-alate and the innovations setose. Although very striking in general aspect, it cannot be matched to any description on the previous revisions. There has been no work on Papuasian *Medinilla* since Merrill and Perry, so it is likely that the species is undescribed.

Poikilogyne aff. *hirta* sensu H. Streimann & P. F. Stevens; colls. 11396 & 11561. A densely setiferous *Poikilogyne* species is locally common in foothill forest near the research house. It does not conform to any of Maxwell's taxa or to annotated material at Lae Herbarium, but comes close to two sheets from Morobe province identified by the collectors (H. Streimann and P. F. Stevens) as *P. hirta*. Unfortunately the binomial is unreferenced and its source cannot be traced with the literature available at the national herbarium. This is possibly an undescribed species or a previously described but rare taxon.

Poikilogyne aff. *villosa* Maxw.; colls. 11343 and 11464. The lax setiform indumentum is reminiscent of *P. villosa* but the Lakekamu populations are not referable to that name. The panicles are subglabrous and much less robust than in *P. villosa*. The leaf hairs are also very sparse and generally restricted to the costa, a situation not found in *villosa*. Possibly a new species or a rare taxon not represented in the collections at Lae Herbarium.

Meliaceae

Dysoxylum phaeotrichum Harms; coll. 11646. Initially misidentified as *Aphanamixis* because of the unusual flagelliform infructescence. The species is rarely collected, being known only from a few sites in central New Guinea (Mabberley *et al.* 1995). Localized endemics of this sort will be the first plants to become endangered in any intensification of logging pressure in Papuan environments.

Monimiaceae

Levieria acuminata (F. v. M.) Perkins; 11382, 11548, and 11566. *Levieria acuminata* is a Queensland species whose range was extended to New Guinea on the basis of 7 specimens assigned to this taxon by Philipson (1980, 1986). All previously cited collections from PNG have originated from Morobe and Central Provinces between 1200 - 3000 m elevation. The Lakekamu record places the species in Gulf Province and suggests that the plant is more common than the small number of collections might indicate. This supposition is also supported by recent collections taken from the Crater Mountain Wildlife Management Area (D. Wright 1341), which have been determined as *L. acuminata* (W. Takeuchi). All previously annotated exsiccatae from PNG are of definitely montane origin, though the Queensland populations are said to descend nearly to sea-level (ibid). Our expedition gatherings from 140 m are substantially below the usual altitude for PNG material, and provide an elevational link between the Australian and Papuasian populations.

Myrsinaceae

Rapanea acrosticta Mez, coll. 11612. The genus *Rapanea* is under revision (J. Pipoly). Malesian taxa formerly in *Rapanea* will be transferred to *Myrsine*. This collection is preliminarily det. as sp. nov. or an aberrant *acroscticta* (Pipoly, personal communication).

Rapanea lamii Sleumer; coll. 11473. Collection 11473 keys to this result on Sleumer (1986) and conforms to the type description. All LAE specimens in the genus are away on loan so the determination cannot be verified against reference material, though the character agreement with published information is very good. *Rapanea lamii* was previously known only from Irian Jaya (ibid). If correctly identified, the Lakekamu occurrence is a first record for Papua New Guinea. Pipoly (personal communication) is uncertain of the identity of *R. lamii* and considers coll. 11473 to represent *Myrsine cruciata*, a species known only from ultramafics from western Malesia. If so, this represents the first record for New Guinea.

Myrtaceae

Syzygium sp. nov.; aff. '*cladopterum-madangense* group'; coll. 11601. A highly unusual species which does not key out on the only available conspectus by Hartley & Perry (1973). The distinctive features of this suspected novelty are the extremely large cordate-based leaves and inflorescences developed as large mounds at ground level. Flowers have the calyx tube cylindriform rather than the usual turbinate-obconical configuration. There is no Papuasian species with such a combination of characters.

Pittosporaceae

Pittosporum ramiflorum (Zoll. & Mor.) Zoll.; coll. 11447. Our RAP specimen (c. 100 m elevation in alluvial forest) keys out very quickly (Bakker & van Steenis 1957). The species is reported to descend to sea level, but usually occurs above 1000 m (ibid). Lae Herbarium has numerous sheets of *P. ramiflorum*, nearly all from the montane zone.

Tectaria group

Chlamydogramme hollrungi (Kuhn) Holttum; coll. 11483. *Chlamydogramme* was erected by Holttum (1986) as a segregate from *Tectaria* sensu stricto. There are two very rare species, both endemic to Mamose region in PNG. *Chlamydogramme hollrungii* was known to Holttum (1991) by just three collections, supposedly from the Sepik region. The only extant populations were believed to be in the Hunstein subdistrict (Sohmer *et al.* 1991, Takeuchi 1994). This species is never in extensive colonies, but here and there in diffuse occurrences, sequestered in dense shade beneath mature canopy. It grows as a terrestrial with fragile stipes easily damaged by traffic through its understory habitat. The Hunstein populations are now at risk due to the impending start of logging operations. Because of the species' ecology and its susceptibility to physical disturbance, the previously known populations are likely to be seriously attrited. Holttum (1991) cites *Hemigramma grandifolia* Copel., a taxon published in 1911, as a synonym of the present species. The *Hemigramma* type consists of a collection by C. King (no. 328) and is said to have originated in NE New Guinea, but since the label reads 'Lakekamu,' it is apparent that the specimen was erroneously attributed to the northern side by Holttum. Copeland (1949, pg. 211) had indicated the existence of 'great confusion in the numbering of King's specimens,' when commenting on King 320, a collection not coincidentally numerically very close to gathering 328. The RAP expedition has almost certainly rediscovered a type population, one of two populations of the species now extant, and the only one thus far known in Papua. Due to endangerment of the Mamose populations, the Lakekamu plants are likely to be of significant conservation value in the future.

Theaceae

Ternstroemia britteniana F.v.M.; coll. 11408. A climbing epiphyte common in hills around the research station. Barker (1980) made wholesale reductions to synonym of Kobuski's taxa, to this particular species, so the binomial covers a range of variation somewhat wider than congeners. However the species is definitely montane in assignment, with a distributional center between 1500 - 3050 m and not reported below 500 m (ibid). The RAP specimen from 137 m is the lowest elevation recorded for the plant. A curious aspect of the Lakekamu population is the epiphytic growth habit, which is not reported in descriptions of this species. *Ternstroemia britteniana* is known as a subcanopy or canopy tree starting from the ground. Barker (ibid) notes the similarity of *T. urdanatensis* of western Malesia to the PNG taxon, and comments that the extra-Papuasian species differs in its epiphytic scandent-shrub habit and the longer pedicels: characters which happen to fit our collection. A further point of uncertainty is the resemblance of no. 11408 to *T. meiocarpa,* a Papuasian montane species closely allied to *T. britteniana* but known only from Irian Jaya. Unfortunately the diagnostic taxonomic distinctions are based entirely on female plants and the RAP vouchers are male. Barker has also expressed reservations regarding the possible conspecificity of the New Guinea taxa, so the binomial assignment is not without qualification. The 'elevational record' may actually be a geographic record of species not previously known from PNG. But with the current state of taxonomic knowledge, it is impossible to be sure.

Tiliaceae

Trichospermum sp. nov.; coll. 11589. Does not key out on any of the leads in Kostermans (1962b). The abaxial surface is laxly provided with erect stellate hairs, and scales are absent. Among Papuasian taxa only *T. tripyxis* has this characteristic, but the latter species has trivalved capsules. The compressed unilocular-bivalved fruits of the present collection indicate affinity to *T. peekelii* but again the indumentum is grossly inconsistent with that species. Our collection also has the twigs with appressed stellate scales rather than the erecto-patent covering on most PNG species. The new species is represented by a complete collection that includes stipules, flowers, and fruits.

Vittariaceae (*Adiantum* group s. lat.)

Vittaria angustifolia Bl.; coll. 11500. A Malesian montane species characterized by Holttum (1954) as an epiphyte from 2,000 - 6,000 ft. elevation. Lae Herbarium has three sheets; all representing gatherings above 700 m. The RAP collection, taken at 175 m, is below the altitudinal range for this species.

Social Hymenoptera Collected During the Survey

TAXA	STRATA[1]	ECOLOGY[2]
APIDAE (SOCIAL BEES)		
Apis cerana Fabricius		
Trigona (Tetragona) genalis Friese		
Trigona (Tetragona) keyensis Friese		
Trigona (Tetragona) sapiens Cockerell		
FORMICIDAE (ANTS)		
Aenictinae		
Aenictus huonicus Wilson	1	G
Aenictus sp. A (nr. *chapmani* Wilson)	1	G
Aenictus sp. B (male)	1	G
Aenictus sp. C (male)	1	G
Aenictus sp. D (male)	1	G
Aenictus sp. E (male)	1	G
Aenictus sp. F (male)	1	G
Cerapachyinae		
Cerapachys desposyne Wilson	1	S
Cerapachys marginatus Emery	1	S
Cerapachys n. sp. A	1	S
Cerapachys n. sp. B	1	S
Cerapachys n. sp. C	1	S
Cerapachys n. sp. D	1	S
Cerapachys n. sp. E	1	S
Cerapachys flavaclavatus Donisthorpe?	1	S
Cerapachys n. sp. G	1	S
Cerapachys n. sp. H	1	S
Dolichoderinae		
Anonychomyrma scrutator (F. Smith)	2	G
Anonychomyrma sp. A	2	G
Anonychomyrma sp. B	2	G
Dolichoderus monoceros Emery	3	G
Iridomyrmex anceps Forel	1	G
Leptomyrmex gracillimus Wheeler	1	G
Leptomyrmex niger Emery	1	G

TAXA	STRATA[1]	ECOLOGY[2]
Philidris sp. A	1,2	G,P
Philidris sp. B	2	G,P
Philidris sp. C	2	P
Philidris sp. D (female)	2	?
Tapinoma melanocephalum (Fabricius)[3]	1	G
Tapinoma sp. A	1	G
Technomyrmex albipes (F. Smith)[3]	2	G
Turneria postomma Shattuck	3	G
Formicinae		
Acropygya acutiventris Roger	1	P
Acropyga sp. A	1	P
Acropyga sp. B	1	P
Acropyga sp. C	1	P
Calomyrmex laevissimus (F. Smith)	2	G
Camponotus dorycus (F. Smith)	1	G
Camponotus papua Emery ?	1,2	G
Camponotus quadriceps (F. Smith)	2	G
Camponotus vitreus (F. Smith)	3	G
Camponotus sp., *conithorax* group	3	G
Camponotus sp. A	?	G
Camponotus sp. B	?	G
Camponotus sp. C	?	G
Camponotus sp. D	2	G
Camponotus sp. E	3	G
Camponotus sp. F	3	G
Camponotus sp. G	?	G
Camponotus sp. H[4]	3	G
Echinopla australis Forel	3	G
Echinopla silvestrii Donisthorpe	3	G
Echinopla n. sp.	3	G
Euprenolepis sp.	1	G
Oecophylla smaragdina (Fabricius)	3	G
Paratrechina longicornis (Latreille)[3]	1	G

TAXA	STRATA[1]	ECOLOGY[2]
Paratrechina minutula Forel	1	G
Paratrechina pallida Donisthorpe	1	G
Paratrechina sp. A	1	G
Paratrechina sp. B	1	G
Paratrechina sp. C	1	G
Paratrechina n. sp. ?	1	G
Polyrhachis (Aulacomyrma) sp.	3?	G
Polyrhachis (Campomyrma) xiphias F. Smith	2	G
Polyrhachis (Chariomyrma) limbata Emery	3	G
Polyrhachis (Chariomyrma) scutulata F. Smith	2	G
Polyrhachis (Chariomyrma) sp. S	2	G
Polyrhachis (Chariomyrma) sp. T	2	G
Polyrhachis (Cyrtomyrma) n. sp. A	2	G
Polyrhachis (Cyrtomyrma) n. sp. B	3?	G
Polyrhachis (Cyrtomyrma) n. sp. C	3?	G
Polyrhachis (Hedomyrma) calliope Emery	2	G
Polyrhachis (Hedomyrma) fervens F. Smith	2	G
Polyrhachis (Hedomyrma) melpomene Emery	2	G
Polyrhachis (Myrma) andromache Roger	3	G
Polyrhachis (Myrma) continua Emery	2	G
Polyrhachis (Myrma) rufofemorata F. Smith	2	G
Polyrhachis (Myrma) sericata (Guerin)	3	G
Polyrhachis (Myrmotopa) bubastes F. Smith	2	G
Polyrhachis (Myrmhopla) greensladei Koh.	2	G
Polyrhachis (Myrmhopla) hortensis Forel	2	G
Polyrhachis (Myrmhopla) nigriceps F. Smith	2	G
Polyrhachis (Myrmhopla) sp., *bicolor* group	2	G
Polyrhachis (Myrmhopla) sp., *sexspinosa* group	2	G
Polyrhachis (Myrmhopla) sp. Q	2	G
Polyrhachis (Myrmhopla) sp. R	2?	G
Polyrhachis (Myrmotopa) alphea F. Smith	2	G
Polyrhachis (P.) bellicosa F. Smith	2	G
Polyrhachis (P.) erosispina Emery	2	G

TAXA	STRATA[1]	ECOLOGY[2]
Pseudolasius sp.	1	G
Myrmicinae		
Adelomyrmex biroi Emery	1	G?
Cardiocondyla nuda Emery[3]	1	G
Cardiocondyla thoracica (F. Smith)	1	G
Cardiocondyla wheeleri Viehmeyer	1	G
Cardiocondyla sp., *paradoxa* group	1	G
Crematogaster paradoxa Emery	2	G
Crematogaster tetracantha Emery ?	2	G
Crematogaster sp. A	1	G
Crematogaster sp. B	3	G
Crematogaster sp. C	3	G
Crematogaster sp. D	?	G
Crematogaster sp. E	?	G
Crematogaster sp. F	?	G
Crematogaster sp. G	?	G
Crematogaster sp. H	?	G
Crematogaster sp. I	?	G
Dacetinops ignotus Taylor	1	?
Dilobocondyla cataulacoidea Stitz	2	?
Eurhopalothrix brevicornis (Emery)	1	S
Eurhopalothrix punctata Szabó	1	S
Eurhopalothrix szentivanyi Taylor	1	S
Eurhopalothrix sp. A	1	S
Lordomyrma n. sp. A	2	G
Lordomyrma n. sp. B	2	G
Lordomyrma n. sp. C	2	G
Mayriella n. sp.	1	?
Meranoplus armatus (F. Smith)	1	G
Meranoplus astericus Donisthorpe	1	G
Metapone sp. A	1	?
Metapone sp. B	1	?
Monomorium floricola (Jerdon)[3]	2	G

TAXA	STRATA[1]	ECOLOGY[2]
Monomorium sp. A	2	G
Monomorium sp. B	1	G
Myrmecina sp. A	1	G
Myrmecina sp. B	1	G
Myrmecina sp. C	1	G
Oligomyrmex atomus Emery	1	G
Oligomyrmex crassiusculus Emery	1	G
Oligomyrmex sp. A	1	G
Pheidole impressiceps Mayr	2	G
Pheidole sp., *sexspinosa* group	2	G
Pheidole sp. A	1	G
Pheidole sp. B	1	G
Pheidole sp. C	1	G
Pheidole sp. D	1	G
Pheidole sp. E	1	G
Pheidole sp. F	1	G
Pheidole sp. G	1	G
Pheidole sp. H	1	G
Pheidole sp. I	1	G
Pheidole sp. J	1	G
Pheidole sp. K	1	G
Pheidologeton affinis (F. Smith)	1	G
Podomyrma gastralis F. Smith	1?	G
Podomyrma sp. A	2	G
Podomyrma n. sp. B	1	P
Podomyrma sp. C	?	?
Podomyrma sp. D	?	?
Podomyrma sp. E (alate female)	?	?
Pristomyrmex sp. A	1	G
Pristomyrmex sp. B	1	G
Pristomyrmex sp. C	1	G
Pristomyrmex sp. D	1	G
Pristomyrmex sp. E	1	G

TAXA	STRATA[1]	ECOLOGY[2]
Pristomyrmex sp. F	1	G
Rhopalothrix diadema Brown & Kempf	1	S
Rhoptromyrmex melleus Forel	1	G
Rogeria stigmatica Emery	1	G
Solenopsis sp. A	1	G
Solenopsis sp. B	1	G
Strumigenys hemichlaena Brown	2	S
Strumigenys horvathi Emery	1	S
Strumigenys loriai Emery	1,2	S
Strumigenys mayri Emery	1	S
Strumigenys sisyrata Brown	?	S
Strumigenys szalayi Emery	1	S
Strumigenys wallacei Emery	1	S
Strumigenys n. sp. (Brown MS)	1	S
Strumigenys sp. A	1	S
Strumigenys sp. B	1	S
Strumigenys sp. C	1	S
Strumigenys sp. D	1	S
Strumigenys sp. E	1	S
Strumigenys sp. F	1	S
Strumigenys sp. G	1	S
Strumigenys sp. H	1	S
Tetramorium fulviceps Emery	1	G
Tetramorium insolens (F. Smith)	1	G
Tetramorium pulchellum Emery	1	G
Tetramorium rigidum Bolton	1	G
Tetramorium validiusculum Emery	1	G
Tetramorium sp. A	1	G
Tetramorium sp. B	1	G
Tetramorium sp. C	1	G
Tetramorium sp. D	1	G
Trichoscapa karawajewi Brown	1	S
Trichoscapa themis (Bolton MS)	1	S

TAXA	STRATA[1]	ECOLOGY[2]
Vollenhovia sp. A	1	G
Vollenhovia sp. B	1	G
Vollenhovia sp. C	1	G
Vollenhovia sp. D	1	G
N. sp., undescr. genus (Taylor MS)	1	?
Incertae sedis	1	?
Ponerinae		
Amblyopone sp.	1	G
Anochetus cato Forel	1	G
Anochetus fricatus Wilson	1	G
Cryptopone butteli Forel	1	G
Cryptopone testacea Emery	1	G
Diacamma rugosum (Le Guillou)	2	G
Discothyrea clavicornis Emery	1	S
Gnamptogenys n. sp.	1	G
Hypoponera confinis (Roger)	1	G
Hypoponera pallidula (Emery)	1	G
Hypoponera papuana (Emery)	1	G
Hypoponera pruinosa (Emery)	1	G
Hypoponera sabronae (Donisthorpe)	1	G
Hypoponera sp. A	1	G
Hypoponera sp. B	1	G
Hypoponera sp. C	1	G
Hypoponera sp. D	1	G
Hypoponera sp. E	1	G
Leptogenys breviceps Viehmeyer	1	S?
Leptogenys diminuta (F. Smith)	1	G
Leptogenys foreli Mann	1	G?
Leptogenys optica Viehmeyer	1	S?
Leptogenys sp. A	1	?
Myopias delta Willey & Brown	1	S
Myopias sp. A	1	S
Myopias sp. B	1	S?

TAXA	STRATA[1]	ECOLOGY[2]
Myopias sp. C	1	S?
Myopias sp. D	1	S?
Myopias sp. E	1	S?
Myopias sp. F	1	S?
Myopias sp. G	1	S?
Myopopone castanea (F. Smith)	1	G
Odontomachus cephalotes F. Smith	1	G
Odontomachus infandus F. Smith?	1	G
Odontomachus saevissimus F. Smith	2	G
Odontomachus simillimus F. Smith	1	G
Odontomachus testaceus Emery	1	G
Odontomachus tyrannicus F. Smith	1	G
Pachycondyla australis (Forel)	1	G
Pachycondyla croceicornis (Emery)	1	G
Pachycondyla papuana Viehmeyer	1	G
Pachycondyla ruficornis Clark	1	G
Pachycondyla stigma (Fabricius)	1	G
Pachycondyla striatula Karavaiev	1	G
Pachycondyla sp. A	1	G
Pachycondyla sp. B	1	G
Platythyrea parallela F. Smith	1	G
Platythyrea quadridentata Donisthorpe	1	G
Ponera szaboi Wilson ?	1	G
Ponera sp. A	1	G
Prionopelta opaca Emery	1	G
Probolomyrmex sp.	1	S?
Proceratium n. sp.	1	S?
Rhytidoponera araneoides (Le Guillou)	1	G
Rhytidoponera inops Emery	1	G
Rhytidoponera nexa Stitz	1	G
Rhytidoponera strigosa Emery	1	G
Pseudomyrmecinae[6]		
Tetraponera laeviceps (F. Smith)	2	G

TAXA	STRATA[1]	ECOLOGY[2]
Tetraponera sp. cf. *laeviceps* (F. Smith)	2	G
Tetraponera modesta (F. Smith)	2	G
Tetraponera nitida (F. Smith)	2	G

VESPIDAE (SOCIAL WASPS)

Polistinae

Parapolybia varia furva Vecht[7]

Polistes bambusae Richards

Polistes tepidus (Fabricius)

Polistes n. sp.

Polistes sp. near *riekii* Richards

Ropalidia deminutiva Cheesman

Ropalidia melania Richards

Ropalidia nigra (F. Smith)

Ropalidia pratti Cheesman

Ropalidia cf. *zonata* Cameron

Ropalidia pratti Cheesman

Ropalidia sp. near *trichophthalma* Richards

Ropalidia conservator (F. Smith)

Ropalidia sp., *deceptor* (F. Smith) group

Ropalidia cf. *brunnea* (F. Smith) (sp. C)

Ropalidia fluviatillis (Meade-Waldo) (sp. D&E)

Ropalidia sp. F

Ropalidia sp. G

Ropalidia sp. H

2*Ropalidia* sp. I

Stenogastrinae

Anischnogaster iridipennis (F. Smith)

Stenogaster concinna Vecht

Vespinae

Vespa trimeres Vecht

[1] The ecological stratification of the ants of the lower Busu River was described by Wilson (1959) in which he recognized three reasonably distinctive strata. In the following list of Ivimka ants, each species is characterized according to this same plan, the number following the name indicating 1-*Ground stratum;* 2-*Low arboreal stratum;* 2-*High arboreal stratum.* See the text for descriptions of strata.

[2] Ants may be further characterized by their general feeding habits. Thus some ants may be specialist predators (S), general predators (G), pastoralists (P), or seed harvesters (H).

[3] Non-native (adventive) in this area.

[4] *Camponotus* sp. H was collected after the RAP survey and is represented by two worker specimens recovered from a malaise trap sample collected by Kurt Merg.

[5] Most *Polyrhachis* species det. by R. J. Kohout.

[6] All *Tetraponera* det. by P. S. Ward.

[7] Previously known only from types collected in the Vogelkop Mts., Irian Jaya. Specimens from other localities in PNG have been seen in the collections of the Bishop Museum, Honolulu.

Natural History Notes on the Social Hymenoptera

Apidae

The social bees at Lakekamu presumably nest in tree hollows high above the ground. No nests were seen. Examination of plants in bloom yielded no additional species of social bees, although one large (up to 2.5 cm long) black species of leaf-cutter bee, *Megachile clotho* (F. Smith), was common at legume flowers along the Avi Avi River. Little can be said of the few species collected at Ivimka beyond the observation that all are forest dwelling species.

Aenictinae

The Aenictinae includes New Guinea's only known true legionary or army ants, all belonging to the genus *Aenictus*. Legionary ants are group raiders that do not have established nests. As far as known they are specialized predators of other ant species (Wilson 1964). Colonies have a single queen and may number into the hundreds of thousands. Foraging columns of *Aenictus* workers were commonly seen during late afternoon as they filed across paths or cleared areas

Cerapachyinae

Cerapachyines are generally considered to be rare; they are certainly unobtrusive and seldom seen. Colonies usually consist of a single queen and a few dozen workers. As far as known, all are predaceous; some, at least, are specialized predators on other ants, invading the nests and taking the brood as prey; others prey on termites. In the New World, *Cerapachys* have been found running in foraging columns of army ants, themselves predators on other ant species.

Dolichoderinae

When crushed, the ants produce a rank, highly repugnant odor, the principle component of which is often butyric acid. Colony defense is further enhanced by the aggressive nature of most of the species. When disturbed, the workers swarm out in large numbers and, while they cannot sting, their bites can be annoying. Colonies are usually populous and seem always to have several queens. Most of our species nest in almost any suitable plant cavity, especially hollow branches, old termite galleries in dead logs, and under epiphyte root mats; at Ivimka the domatia of the myrmecophyte, *Myrmecodia,* were almost exclusively monopolized by *Anonychomyrma scrutator*. The species of *Anonychomyrma* are among the most often encountered of our ants: the workers forage on vegetation in long, dense files running up the trunks of trees and shrubs and across vines. The several species of both *Anonychomyrma* and *Philidris* are general predator-scavengers and pastoralists that tend aphids and mealybugs on various shrubs and trees. These Homoptera are often hidden under small shelters of plant fibers on the host plant. The two species of *Leptomyrmex* are aggressive predators and usually nest in soil or in rotten wood on or near the ground

Formicinae

Formicinae are stingless; they rely almost wholly upon chemical defense. When agitated they spray concentrated formic acid (so-named because it was originally derived from the Palearctic ants of the genus *Formica*). Formic acid is a powerful insecticidal and fungicidal fumigant. Some of the smaller species, especially those in genera such as *Acropyga* and *Paratrechina* rely more upon their small size and cryptic habits. One species of *Camponotus, C. dorycus,* is the largest ant at Ivimka; colonies are populous and, when disturbed, the ants aggressively defend the colony; the large workers (so-called soldiers) can draw blood when they bite. They add insult to injury when they spray formic acid into the wound.

Acropyga and *Paratrechina* nest in decaying wood, as do some of the *Camponotus* and *Polyrhachis*. Other species of *Camponotus* nest in living plant stems, where they subsist largely on the exudates of homopterans that feed on the plants from within the stems. Many species of *Polyrhachis* build nests of masticated plant fibers (carton) or felted plant hairs (felt) that are attached to large leaves of palms or *Pandanus*. A few *Polyrhachis*, and *Oecophylla smaragdina*, build arboreal nests of living leaves joined together by the woven silk of the ant larvae.

Formicinae are mostly generalized predator-scavengers, but with a strong predilection for plant-derived carbohydrates, either in the form of nectar from flowers or extra-floral nectaries or as secretions ("honey-dew") from aphids, mealybugs, and other Homoptera. Larvae of certain families of butterflies (Lepidoptera: Lycaenidae and Noctuidae) are assiduously tended and protected by formicine ants, especially *O. smaragdina;* the ants thus protect herbivores that may do considerable localized damage to individual trees and shrubs.

Oecophylla smaragdina ranges from India to Australia, east to Taiwan. This "weaver ant" is well-known as an unusually aggressive predator that builds multidomous arboreal nests that may occupy several trees. In much of the range of this species in eastern Australia, the ants are green, rather than red as in our populations; these are the "green ants" of the Aboriginal "Dream Time".

Myrmicinae

While most species of Myrmicinae can sting, they are mostly too small to be painful. Only the species of *Podomyrma* are capable of an annoying sting. Other Myrmicines, however, respond to disturbance in sufficiently large numbers as to be at least moderately annoying.

While some species, especially in the genera *Crematogaster, Lordomyrma,* and *Podomyrma* are arboreal, nesting in hollow stems and branches, most of the myrmicines nest in soil and litter or in rotting wood. Some of the species of *Strumigenys* nest under bark chips on *Ficus* trunks, while others nest in rotting wood.

Feeding habits are diverse: most Myrmicinae are generalist predator-scavengers and aggressively recruit to newly-discovered food resources and drive other ants away. Some genera, such as *Strumigenys* and *Trichoscapa* are specialized predators, usually on minute litter-dwelling arthropods. Various species of *Pheidole* are probably seed-gatherers at Ivimka, as they are elsewhere in the world; some, however, are probably general scavenger-predators. We assume such genera as *Rhopalothrix, Eurhopalothrix, Dacetinops,* and *Pristomyrmex* to be specialized predators also, but these are so poorly known that little can be said of them.

Cardiocondyla nuda and *Monomorium floricola* are exotic species that probably have little or no deleterious impact on the native species. Although originally described from Fiji, *C. nuda* is assumed to be African. Since the ant was found only near the Avi Avi River, but well removed from the camp area, it seems likely that this does not represent a recent introduction. Perhaps *C. nuda* was brought into the area during the Second World War on supplies being carried to Wau via the Bulldog Road.

Presumably originally a tropical Asian species, *M. floricola* is now a widespread tropicopolitan "tramp" species. According to Wilson and Taylor (1967), this ant is "almost wholly arboreal, forming large colonies in trees and bushes in habitats of various degrees of disturbance. It is a prominent urban species in most tropical countries. Colonies seem unable to penetrate undisturbed native forests." While the latter point is debatable, it generally seems to be true. At Ivimka, *M. floricola* is not a commonly encountered species and was found only in the temporary camp. It is possible that it was brought in recently on supplies carried in from Port Moresby.

Ponerinae

Ponerine ants are largely predaceous and a few are evidently specialized predators. Several species of *Myopias* are specialists on millipedes (Myriapoda) (Willey and Brown 1983; personal observation). Others, such as species of *Discothyrea* and *Proceratium* (Brown 1958) and of *Probolomyrmex* (Taylor 1965) are suspected of being predators on eggs of various litter inhabiting arthropods. Most ponerines are solitary foragers, but some, such as *Odontomachus,* recruit very rapidly to a concentrated prey supply, such as exposed termites in a rotten log. The species of *Leptogenys* are nomadic and are group raiders, at least some of which (*L. breviceps*) prey largely on termites. On several occasions, workers of *Rhytidoponera nexa* were seen carrying moribund leeches; whether the leeches had been actively attacked is uncertain.

Most of our species nest in soil and/or leaf litter, often under a covering object such as a stone or piece of wood, or in rotting wood. A few, such as *Diacamma* and *Odontomachus* may be opportunistically "arboreal", nesting in preformed cavities in dead tree branches or dense root mats of epiphytes, usually within 5 meters of the ground.

Ponerines characteristically are considered to be stinging ants and some of our species, especially in the genera *Leptogenys, Odontomachus,* and larger *Pachycondyla* are capable of inflicting quite painful stings. Others of similar size, such as the larger *Myopias* and *Rhytidoponera,* have surprisingly mild stings.

Pseudomyrmecinae

There is only one Old World genus (*Tetraponera*), of which all are arboreal and all are largely predaceous on other arthropods. Most species nest in hollow twigs and/or branches of trees and shrubs; a few species are obligate residents of Asian bamboo species. While some of the largest species are capable of delivering a painful sting, most species are timid and innocuous.

Vespidae (Social Wasps)

All of the social wasps are predators on other arthropods. Polistinae are mainly predaceous on larval Lepidoptera (caterpillars), but occasionally will take other prey. The one vespine genus at Lakekamu, *Vespa,* attacks the nests of other social Hymenoptera, especially the various Polistinae; the adult wasps are driven away from the nest; the *Vespa* worker then removes the larvae and pupae. Although Stenogastrinae are predatory, little detailed information is available. The hunting wasps are most often seen flying slowly about in dark and secluded areas. Some species are known to prey on chironomid midges (Diptera: Chironomidae) and on spiders.

Polistes is a cosmopolitan genus that is best represented in the Afrotropical and in the Neotropical regions. The *Polistes fauna* of New Guinea is not well studied and the number of species present is uncertain but probably there are about 12-15 species. Colonies usually are rather small, seldom exceeding 50 individuals. The nest consists of a single comb suspended from a branch or other supporting object and is never enclosed within an envelope. Lepidoptera larvae are the usual prey but the foraging workers will often attack other relatively soft-bodied insects. The wasps are generally not especially aggressive but are capable of delivering a painful sting.

One of our species is *P. tepidus,* a large black species with rich yellow markings was originally described from Australia. Vecht (1971) treated the New Guinean form as a separate subspecies, *P. tepidus malayanus* Cameron, which, despite its name, was described from Manokwari, Irian Jaya. Pending a detailed study of more material from both regions I have opted to ignore any attempt at subspecific recognition. Another of the Ivimka species is *P. bambusae,* described from Wau, nests in the hollow stems of large bamboo (Richards 1978). I collected only two foraging workers. Two additional species remain unidentifiable at this time. One of these is apparently similar to *P. laevigatissimus* G. Soika (known only from the types from Broome, West Australia) and the other similar to *P. riekii* Richards, described from the Cape York Peninsula, Queensland.

Ropalidia is not only the most abundantly represented social wasp genus, it is the most diverse in nesting habits. Some species construct small, single comb nests that lack an envelope under leaves or tree branches. Others (*R. pratti* and *R.* cf. *pratti*) construct single comb nests, again lacking an envelope, in hollow logs or tree trunks or cavities in soil. At least one species (*R.* cf. *zonata*) constructs multiple comb nests that lack an envelope. Another species, *R. deminutiva,* attaches its enveloped, single comb nest to the underside of large leaves. A football-size multiple comb nest that is enclosed by an envelope is made by *Ropalidia fluviatillis;* these nests are found hanging on shrubs or low branches. Mature nests of this species may contain several hundred workers.

Stenogastrinae are rare, secretive wasps that are seldom collected. The few species are primitively social. Several females may be present on a nest, but each constructs her own cells and is responsible for the larvae that develop therein. The nests are very simple: a single series of cells arranged in linear fashion along a thread-like stalk. Nests are situated in dark locations, such as the overhangs along river banks, in caves, *etc.,* where they are concealed among hanging rootlets. The wasps are timid and make no effort to defend the nests.

The subfamily Vespinae reaches its southern terminus in New Guinea, where there are two species of *Vespa, V. affinis* (Linné) and *V. trimeres* Vecht. At Ivimka, *V. trimeres* was present, but seldom encountered. Little is known of the ecology of this ground-nesting species, but it is presumably a predator on other social Hymenoptera, a trait common to many species of *Vespa* (Spradbery 1973; Starr 1992; Vecht 1957). Several polistine nests with only very small larval cells were seen that were occupied by a queen and up to six workers; such nests apparently are initiated by the survivors of a nest depredated by a foraging *Vespa* worker. I observed other polistine nests in Hong Kong that had been attacked by foraging *Vespa:* the polistines made no effort to defend the nest, but abandoned it to the much larger hornet, which proceeded to remove the larvae. The survivors initiated a new nest at a nearby location.

Foraging Vespa workers were occasionally seen flying slowly through dense vegetation near ground level, apparently searching for prey. One nest was located in the ground at the base of a tree, but was not collected.

Odonata Collected During the Survey

Species of Odonata collected in the Lakekamu Basin and their habitat preferences. See text for detailed description of habitat types.

SPECIES	RIVER	STREAM	CLEARING	POND	FOREST
ANISOPTERA					
Agrionoptera longitudinalis biserialis	+		+		
Camacinia othello					+
Diplacina arsinoe	+				
D. erigone			+		
Diplacodes haematodes	+				
Gynacantha mocsaryi				+	
Huonia arborophila arborophila	+				
Ictinogomphus lieftincki		+			
Macromia celaeno			+		
M. eurynome			+		
Nannophlebia amaryllis?	+				
Neurothemis ramburi papuensis			+		
Orthetrum glaucum	+		+		
O. villosovittatum villosovittatum	+	+	+	+	
Protorthemis coronata			+		
Rhyothemis princeps irene	+		+		
Synthemis primigenia			+		
ZYGOPTERA					
Argiolestes montivagans		+			+
Drepanosticta bicornuta?		+			+
D. dendrolagina		+			+
Idiocnemis pruinescens		+			+
Idiocnemis sp. nr *pruinescens*		+			+
Idiocnemis sp. nr *kimminsi*		+			+
Neurobasis australis	+				
Nososticta dorsonigra					+
N. sp. nr *finisterrae*		+			
N. nigrifrons					+

SPECIES	RIVER	STREAM	CLEARING	POND	FOREST
N. nigrofasciata					+
N. rosea?		+			
N. salomonis	+	+			
Papuagrion occipitalis					+
Rhinocypha tincta semitincta	+	+			
Teinobasis angusticlavia					+
Teinobasis? sp. nov.	+				

Species Accounts of Odonata

Anisoptera (Dragonflies)

Agrionoptera longitudinalis biserialis. Widespread throughout New Guinea. Found in sunny clearings and along the Sapoi River

Diplacina arsinoe. Widespread in New Guinea. Found in sunny areas along the banks of the Sapoi River

Diplacina erigone. A taxon collected closely fits the description of this species, a species previously described and known only from Misool Island. The Lakekamu record represents a major range extension or an undescribed species. Found in sunny clearings.

Diplacodes haematodes. Common and widespread in New Guinea and Australia, occupies a wide range of water bodies. At Lakekamu it was common in sunny areas along the Sapoi River.

Camacinia othello. Common, widespread species. A large and beautiful dragonfly with strongly patterned wings. The single specimen collected was perched among dense foliage 2 m above the ground during a violent storm.

Gynacantha mocsaryi. Common, widespread species. Like other members of this genus *G. mocsaryi* is a crepuscular species. Specimens were collected at dusk or in gloomy conditions, flying over the small pond south of Ivimka Camp.

Huonia arborophila arborophila. Widespread species. Found perched on rocks and vegetation on the banks of the Sapoi River.

Ictinogomphus lieftincki. Widespread in New Guinea. One *I. lieftincki* (presumably the same individual) was observed on most days, patrolling a small creek with extensive light penetration due to a large treefall. *Ictinogomphus lieftincki* is the only gomphid occurring on mainland New Guinea.

Macromia celaeno, M. eurynome. Both species were found in forest clearings. Several specimens of each were collected during late afternoon in the camp clearing. They appeared to be crepuscular, being most active between late afternoon and dusk.

Nannophlebia amaryllis. A species distributed mainly in northern New Guinea. Perched in sun along the banks of the Sapoi River.

Neurothemis ramburi papuensis. A widespread species but rather uncommon at Lakekamu. Several specimens were observed in sunny clearings adjacent to camp.

Orthetrum glaucum. A common and widespread species. At Lakekamu found in forest clearings and along the banks of the Sapoi River.

Orthetrum villosovittatum villosovittatum. A common and widespread species, occupying a range of water bodies and also seen in forest clearings near the Sapoi River.

Protorthemis coronata. A widespread species. One of the most conspicuous dragonflies at Lakekamu. This beautiful species was common in forest clearings, and frequently perched on tents and other buildings at camp.

Rhyothemis princeps irene. A common and widespread species. At Lakekamu many individuals were observed hovering over forest clearings and over wide, dry and sunny stretches of the river bank.

Synthemis primigenia. A widespread species found in clearings around the IRS and helipad.

Zygoptera (Damselflies)

Argiolestes montivagans? Known from northeastern New Guinea. The status of the Lakekamu specimens is unclear, as they have a different thoracic pattern to *A. montivagans*. They may be a variety of *A. montivagans* or an undescribed species.

Drepanosticta bicornuta? The status of these specimens is unclear. If they represent *D. bicornuta* this record is a large range extension from northwestern New Guinea. At Lakekamu found along small streams and perched in the forest understory.

Drepanosticta dendrolagina. Previously known from northeastern New Guinea. At Lakekamu found along small shaded streams and perched in the forest understory.

Idiocnemis pruinescens. Found in eastern New Guinea, this species was abundant along trails and small shaded streams.

Idiocnemis sp. nr *pruinescens.* An undescribed species that is morphologically similar to *I. pruinescens* but has different anal appendages.

Idiocnemis sp. nr *kimminsi.* An undescribed species similar to *I. kimminsi* but having different anal appendages.

Neurobasis australis. Widespread. A large and conspicuous species, common in sunny areas along the Sapoi River and frequently engaging in elaborate courtship displays over water. One male was attacked and knocked to the water by a large dragonfly during display.

Nososticta dorsonigra. A series of specimens most like *N. dorsonigra,* a species known from the Vogelkop Peninsula. At Lakekamu found in the forest understory.

Nososticta nr *finisterrae.* Almost certainly an undescribed species related to *N. finisterrae.* At Lakekamu found only along small, shady creeks.

Nososticta nigrifrons. Widespread in southern New Guinea, at Lakekamu this species was found on low vegetation in the forest understory.

Nososticta nigrofasciata. A widespread species in southern New Guinea, at Lakekamu it was found on vegetation near the forest floor.

Nososticta rosea? Almost certainly an undescribed species related to *N. rosea.* At Lakekamu found perched in dappled sun along small streams.

Nososticta salomonis. A widespread species, at Lakekamu found along small streams in dappled sun and also abundant along sunny, slow-flowing stretches of the Sapoi River 6 km south of Ivimka Camp.

Papuagrion occipitale (light form). Widespread in New Guinea but poorly known. This species was found in dappled sun along forest trails and near the edges of clearings.

Rhinocypha tincta semitincta. Common, widespread. Abundant in sunny patches along the Sapoi River and in a small creek where a treefall increased light penetration. Males frequently engaged in complex and extended aerial displays.

Teinobasis angusticlavia. Previously known only from the Aru Islands. Rather patchily distributed in the forest, this species was found only in dense vegetation within about 15m of the banks of the Sapoi River. They were never found on the river banks, preferring dark, shaded areas. All specimens collected were perched on vegetation between one and two meters above the ground

Teinobasis (?) sp. nov. An undescribed species of uncertain affinities, only tentatively placed in this genus. This species was also recently collected by D. Polhemus in the Kikori Basin and in southern Irian Jaya (personal communication).

APPENDIX 9 Summary of Fish Collecting Sites

STA.	DATE	LOCALITY	APPROX. DISTANCE FROM SEA (km)	ELEVATION (m)	NO. SPECIES
1	11/11/96	Ivimka Creek near camp	95	120	6
2	12/11/96	Sapoi R. at Bulldog crossing	95	130	9
3	13/11/96	Sapoi R. near Avi Avi "junc."	96	140	6
4	14/11/96	Creek on Kakoro Track	100	120	9
5	15/11/96	Sapoi R. - 4.5 km from camp	92	40	11
6	16/11/96	Sapoi R. - side tributary	93	80	8
7	17/11/96	Sapoi R. - 5.5 km from camp	90	35	15
8	17/11/96	Sapoi R. - "Turtle Pool"	91	40	13
9	18/11/96	Avi Avi R. - 5.2 km from camp	90	35	10
10	18/11/96	forest trib. of Avi Avi River	92	40	12
11	19/11/96	Sapoi R. - 2 km upstream of Ivimka Camp	97	200	5
12	20/11/96	forest stream on Kakoro Track, 1 km SE of Ivimka Camp	96	120	8

Annotated Checklist of Fish Collected During the Survey APPENDIX 10

The phylogenetic sequence of the families appearing in this list follow the system that is used by the major Australian museums and approximates that proposed in *Nelson's Fishes of the World* (2nd edition, 1984, John Wiley and Sons). Genera and species are arranged alphabetically within each family.

Text for each species includes a series of annotations, each separated by a semicolon. These annotations pertain to general habitat, detailed habitat, known altitudinal range, general activity mode, social behavior, major feeding type, food items, reproductive mode, maximum size, general distributional range, and additional comments pertinent to the present survey. The length is given as standard length (SL) for most species, which is the distance from the tip of the snout to the base of the caudal fin. Total length (TL) is given for a few fishes which do not have a clearly defined caudal fin (eels and plotosid catfishes for example).

Anguillidae - Freshwater Eels

Anguilla bicolor (McClelland, 1844) - Indian Short-finned Eel
Fresh water; creeks and rivers; up to at least 1000 m elevation; rests on bottom; solitary; carnivore; fishes, crustaceans; spawns pelagic eggs; 60 cm SL; Indo-west Pacific, most records from PNG are from northern locations, particularly the Ramu and Sepik drainages; only a single specimen collected at Lakekamu from a forest stream (Ivimka Creek), but reported by Tekadu villagers as being common and ascending the Avi Avi River beyond Anandea.

Ariidae - Forktail Catfishes

Arius leptaspis (Bleeker, 1862) - Triangular Shield Catfish
Mangroves, tidal creeks, and lowland streams; below about 30-40 m elevation; diurnal benthic; solitary or in groups; omnivore; insects, fishes, molluscs, and plants; male broods eggs in mouth; 50 cm SL; Southern New Guinea and Northern Australia; previously known in PNG in major drainage systems west of the Purari River, the Lakekamu record represents a eastward extension of the range; relatively uncommon in the survey area, it was seen on two occasions - two separate groups of 5 and 15 individuals were sighted in deep pools in the Sapoi River, between 5 and 6 km downstream from Ivimka Camp.

Plotosidae - Eel-tailed Catfish

Neosilurus ater (Perugia, 1894) - Narrow-fronted Tandan
Fresh water; creeks and rivers; nocturnal benthic; solitary or in groups; carnivore; insects, crustaceans, molluscs, worms, and fishes; demersal eggs with no parental care; to about 50 cm TL; Southern New Guinea and Northern Australia; most records from mainland New Guinea are from the Fly Delta, the only previous record from the eastern side of the Gulf of Papua was from Inawi (about 75 km northwest of Port Moresby); uncommon in the study area, but about 10 specimens collected with rotenone from a deep pool of the Sapoi River approximately 6 km downstream from Ivimka Camp.

Neosilurus brevidorsalis (Gunther, 1867) - Short-finned Tandan
Fresh water; creeks and rivers; below about 500 m elevation; diurnal benthic; solitary or in groups; carnivore; insects, crustaceans, molluscs, worms, and fishes; demersal eggs with no parental care; 17 cm SL; Southern New Guinea and Northern Australia; widely distributed in New Guinea from the Vogelkop Peninsula eastward to the Kemp Welsh River near Port Moresby; common throughout the survey area in both forest creeks and larger rivers; according to Tekadu residents it ascends the Avi Avi river as far as Anandea.

Hemiramphidae - Halfbeaks or Garfishes

Zenarchopterus novaeguineae (Weber, 1913) - Fly River Garfish

Fresh water; lowland creeks and rivers; about 20-100 m elevation; surface swimmer; forms aggregations; carnivore; floating insects; eggs spawned on weed or floating debris; 17 cm SL; Southern New Guinea; previously recorded from the Lorentz (Irian Jaya), Fly, and Laloki river systems and therefore presumed to be widespread; less than 10 specimens were seen during the survey near the surface of two deep pools in the Sapoi River, about 6-7 km downstream from Ivimka Camp. Specimens from the study area had yellow fins, a feature not previously noted.

Melanotaeniidae - Rainbowfishes

Melanotaenia goldiei (Macleay, 1883) - Goldie River Rainbowfish

Fresh water; creeks and rivers; to about 500 m elevation; midwater diurnal; forms aggregations; omnivore; insects and their larvae, crustaceans, plants; eggs spawned on weed or floating debris- spawning detected during survey; 11.5 cm SL, females to 10 cm SL; widespread in Southern New Guinea from Lake Yamur (Irian Jaya) eastward to the Port Moresby region; very abundant and the most common fish throughout the survey area, occurring in both forest creeks and rivers; it is most abundant in deep pools, where it forms loose midwater aggregations; the farthest upstream penetration in the Sapoi River was approximately 2 km upstream from Ivimka Camp at the base of formidable cascades.

Melanotaenia splendida rubrostriata (Ramsay and Ogilby, 1886) - Red-striped Rainbowfish

Fresh water; creeks and rivers; below about 450 m elevation; midwater diurnal; forms aggregations; omnivore; insects and their larvae, crustaceans, plants; eggs spawned on weed or floating debris; 12.5 cm SL, females to 10 cm SL; widely distributed in Southern New Guinea west of Etna Bay (Irian Jaya); the Kikori River was the previous eastern limit of distribution, but specimens from the current survey indicate it is more widespread; very rare in the study area, only two adult males collected in deep pools of the Sapoi River, about 5-6 km downstream from Ivimka Camp.

Melanotaenia species - Forest Rainbowfish

Fresh water; creeks and rivers; to 140 m elevation; midwater diurnal; forms aggregations; omnivore; insects and their larvae, crustaceans, plants; eggs spawned on weed or floating debris - spawning detected during survey; 6.5 cm SL; a possible new species with unknown distributional limits; closely related to and possibly just a color variety of *M. papuae;* common throughout the survey area, mainly in slow-flowing forest creeks, but also found in side channels and quiet pools of the Sapoi River in sections 4-7 km downstream from Ivimka Camp. first time.

Atherinidae - Silversides

Craterocephalus randi (Nichols and Raven, 1934) - Kubuna Hardyhead

Fresh water; creeks and rivers; diurnal benthic; forms aggregations; omnivore; algae, microcrustaceans, insects; eggs spawned on weed or floating debris; 8 cm SL; widespread in southern New Guinea between Etna Bay (Irian Jaya) and the Kubuna River (about 80 km northwest of Port Moresby); abundant in Sapoi River, mainly in quiet section downstream from Ivimka Camp.

Terapontidae - Grunters

Hephaestus trimaculatus (Macleay, 1884) - Threespot Grunter

Fresh water; creeks and rivers; roving predator; solitary or in groups; omnivore; algae, insects, crustaceans, molluscs, fishes, and frogs; demersal eggs with no parental care; 22 cm SL; Southern New Guinea, previously known only from the Port Moresby region in the Laloki River and its tributaries; common in deeper pools of the Sapoi River, also found in forest creeks, but in smaller numbers; the farthest upstream penetration in the Sapoi River was approximately 2 km upstream from Ivimka Camp at the base of formidable cascades.

Kuhliidae - Flagtails

Kuhlia marginata (Cuvier, 1829) - Spotted Flagtail

Fresh water; creeks and rivers; 0-5 m; diurnal midwater; forms aggregations; carnivore; insects and larvae, crustaceans, and fishes; spawns pelagic eggs; 18 cm SL; Western Pacific, from New Guinea northward to Japan and eastward to the Society Islands; most previous records from PNG are from north coast drainages; not seen in large numbers, but several individuals noted in most pools at the base of rapids in the Sapoi River; the species penetrates well upstream into mountainous terrain.

Apogonidae - Cardinalfishes

Glossamia sandei (Weber, 1908) - Sande's Mouth Almighty

Fresh water; creeks and rivers; hovers in midwater; solitary; carnivore; fishes and crustaceans; male broods eggs in mouth - ripe females and egg brooding males were collected during the survey; 19 cm SL; widespread in southern New Guinea from Lake Yamur (Irian Jaya) eastward to the Lakekamu Basin; the former eastern boundary of its distribution was the Purari River; uncommon, but about 10 individuals collected during the survey from pools formed by log jams, both in forest creeks and the Sapoi River.

Cichlidae - Cichlids

Oreochromis mossambica (Peters, 1852) - Tilapia

Fresh water; creeks, rivers, and lakes; diurnal benthic; solitary or in groups; herbivore; algae; parental care of demersal eggs; 30 cm SL; native to African fresh waters, but widely introduced throughout the tropics, apparently PNG stock originated from Malaya and was introduced widely in PNG as a source of food by the Department of Stock, Agriculture, and Fisheries beginning in 1954; uncommon in the Upper Lakekamu Basin, only 3 juvenile and subadult specimens were sighted or collected from the Sapoi River; the farthest upstream penetration was approximately 4 km downstream from Ivimka Camp.

Mugilidae - Mullets

Cestraeus goldiei (Macleay, 1884) - Goldie River Mullet

Fresh water; creeks and rivers; diurnal midwater; solitary or in groups; herbivore; algae; demersal eggs with no parental care; 41 cm SL; known from Sulawesi, Timor, New Guinea, Vanuatu, and New Caledonia; most records from PNG are from the Port Moresby district; has distinctive torpedo shape and is adept at penetrating well inland to an altitude of about 350 m in fast-flowing streams; not seen in large numbers, but several individuals noted in most pools at the base of rapids in the main Sapoi River; the species penetrates well upstream into mountainous terrain; in the Avi Avi River it reaches at least as far as Tekadu.

Crenimugil heterocheilus (Bleeker, 1855) - Fringe-lipped Mullet

Fresh water; lowland creeks and rivers; below about 50 m elevation; forms benthic grazing schools; forms aggregations; herbivore; bottom detritus and plants; spawns pelagic eggs; 50 cm SL; Coastal streams of Indonesia, New Guinea, Solomon Islands, and Vanuatu; there are few reports from New Guinea, but it is known from the Fly system and the Kemp Welsh River east of Port Moresby; uncommon in the survey area - but four specimens collected and several small schools seen in the Sapoi River; usually associated with quiet sections of the river, the maximum upstream penetration appears to be situated about 3 km downstream from Ivimka Camp.

Gobiidae - Gobies

Awaous acritosus (Watson, 1994) - Roman-nosed Goby

Fresh water; creeks and rivers; rests on bottom, buries in sand when disturbed; solitary or in groups; omnivore; algae and small crustaceans; parental care of demersal eggs; 16 cm SL; Southern New Guinea and Northern Australia, previously known in PNG only from the Port Moresby region, common in the Sapoi River, increasing in abundance downstream from Ivimka Camp; the Bulldog Track crossing of the Sapoi River just upstream from Ivimka Camp appears to be the limit of upstream penetration; also found in forest streams, but relatively rare in this habitat.

Glossogobius celebius (Valenciennes, 1837)- Celebes Goby

Fresh water; creeks and rivers, usually within a few kilometers of the sea, but sometimes penetrating well inland; rests on bottom; solitary; carnivore; crustaceans and small fishes; parental care of demersal eggs; 12 cm SL; widely distributed in western Pacific including New Guinea, Solomon Islands, northern Australia, Indonesia, Philippines, Taiwan, and Ryukyu Islands - most records from PNG are from northern coastal locations; rare, a single specimen collected in the Avi Avi River.

Glossogobius species - Spotfin Goby

Fresh water; creeks and rivers; rests on bottom; solitary; carnivore; crustaceans and small fishes; parental care of demersal eggs; 5.5 cm SL; an undescribed species with uncertain distributional limits; possibly synonymous with an undescribed species from the Upper Fly River (*Glossogobius* species 9 in Allen, 1991); common on cobble bottoms in the Sapoi River, penetrating well upstream into mountainous terrain; rare in forest creeks.

Lentipes watsoni (Allen, 1997) - Clinging Goby

Fresh water; fast-flowing creeks and rivers; diurnal benthic; solitary or in groups, seen clinging to rocks with its pelvic "sucker" in areas of swift current; omnivore; algae and micro-invertebrates; parental care of demersal eggs; about 8 cm; an undescribed species collected for the first time during the Lakekamu survey and named in early 1997 (Allen 1997); the genus *Lentipes* was unknown from New Guinea previous to 1995, when another undescribed species was found in a drainage of the Cyclops Range near Jayapura; this is the first member of the subfamily Sicydiinae to be recorded from southern New Guinea, but the group is well represented in north coastal streams by the genera *Stiphodon, Sicyopus,* and *Sicyopterus;* common in rapids upstream from Ivimka Camp; in the Avi Avi it penetrates well upstream, at least as far as Anandea; pronounced sexual dichromatism with brightly colored males being outnumbered by females by an approximate ratio of 15 to 1.

Eleotridae - Gudgeons

Mogurnda pulchra (Horsthemke and Staeck, 1990) - Moresby Mogurnda
Fresh water; swamps and creeks; hovers in midwater; solitary; carnivore; insects, crustaceans, and small fishes; parental care of demersal eggs; 11.8 cm SL; Southern PNG between the Laloki River (near Port Moresby) and the Purari River; commonly collected in forest creeks and swamp habitat during the survey, less abundant in the Sapoi River, where it is restricted to quiet sections near shore and the edge of deep pools; previously reported to reach a maximum standard length of 8 cm, but the largest Ivimka specimen measured 11.8 cm SL.

Oxyeleotris fimbriata (Weber, 1908) - Fimbriate Gudgeon
Fresh water; creeks, rivers, and lakes; 10-1500 m elevation; rests on bottom; solitary; carnivore; insects, molluscs, crustaceans, and fishes; parental care of demersal eggs; 16 cm SL; widespread throughout New Guinea, one of the few species occurring on both sides of the Central Dividing Range, also found in Northern Australia on Cape York Peninsula; regularly collected from forest creeks and the Sapoi River; most of the survey specimens are small juvenile - adults are often difficult to collect due to their resistance to rotenone and cryptic habits; penetrates well upstream into the mountains - in the Avi Avi River it is known upstream of Anandea.

Oxyeleotris gyrinoides (Bleeker, 1853) - Greenback Gauvina
Fresh water; lowland creeks and rivers; below about 150 m elevation; rests on soft mud bottoms; solitary; carnivore; insects, molluscs, crustaceans, and fishes; parental care of demersal eggs; 30 cm SL; Indo-west Pacific from Sri Lanka to Pohnpei (Caroline Islands), previously known from New Guinea only from Waigeo (Irian Jaya) and the Ramu and Gogol river systems; a single subadult collected from Ivimka Creek in forest habitat.

Soleidae - Soles

Synaptura villosa (Weber, 1908) - Velvety Sole
Fresh water; creeks and rivers; diurnal benthic on sandy or mud bottoms; solitary; carnivore; small invertebrates; spawns pelagic eggs; 10.5 cm SL; Southern New Guinea between the Lorentz River (Irian Jaya) and the Lakekamu Basin; prior to this survey the Fly River system represented the eastern limit of distribution; relatively common in the Sapoi River, penetrating to the Bulldog Road Crossing a short distance (about 300 m) upstream from Ivimka Camp.

Summary of Fish Collected

GENUS AND SPECIES	STATION NUMBER											
	1	2	3	4	5	6	7	8	9	10	11	12
Anguilla bicolor	X											
Arius leptaspis								X				
Neosilurus ater							X					
Neosilurus brevidorsalis		X	X	X	X	X	X	X	X	X	X	X
Zenarchopterus novaeguineae							X					
Melanotaenia goldiei	X	X	X	X	X	X	X	X	X	X		X
Melanotaenia splendida rubrostriata							X	X				
Melanotaenia sp.	X				X	X	X	X		X		
Craterocephalus randi		X		X	X		X	X	X	X		
Hephaestus trimaculatus		X	X	X	X		X	X	X	X		X
Kuhlia marginata		X		X	X		X	X			X	
Glossamia sandei			X			X	X			X		X
Oreochromis mossambica					X			X				
Cestraeus goldiei		X		X				X				
Crenomugil heterocheilus							X			X		
Awaous acritosus		X			X	X	X	X	X	X		X
Glossogobius celebius									X			
Glossogobius sp.			X	X	X	X	X	X	X	X	X	X
Lentipes watsoni		X									X	
Mogurnda pulchra	X		X	X	X	X	X		X	X		X
Oxyeleotris fimbriata	X		X			X			X	X	X	X
Oxyeleotris gyrinoides	X											
Synaptura villosa		X			X		X	X	X	X		
Total species	6	9	6	9	11	8	15	13	10	12	5	8

CONSERVATION INTERNATIONAL   **Rapid Assessment Program**

Comparison of the Fish Faunas of Major River Systems in New Guinea

Comparison of the fish fauna of various river systems in New Guinea. Data taken from Roberts (1978), Allen (this report), Weber (1913), Haines (1979), Allen and Coates (1990), Allen, Parenti, and Coates (1992), Allen and Boeseman (1982), and Parenti and Allen (1991).

RIVER SYSTEM	TOTAL SPECIES	ENDEMIC SPECIES	PERCENT ENDEMICS
Fly	103	5	4.8
Kikori	100	14	14.0
Lorentz	60	2	3.3
Purari R	5	6	10.5
Sepik	57	0	---
Ramu	54	0	---
Mamberamo*	40	4	10.0
Digul*	40	0	---
Gogol	25	0	---
Lakekamu *	23	?	?

* indicates incompletely surveyed

Amphibians Recorded During the Survey

Myobatrachidae
Lechriodus melanopyga (Doria, 1874)

Ranidae
Rana arfaki Meyer, 1874
Rana daemeli (Steindachner, 1868)
Rana garritor Menzies, 1987
Rana grisea van Kampen, 1913

Hylidae
Litoria dorsalis (Macleay, 1878)
Litoria genimaculata (Horst, 1883)
Litoria graminea (Boulenger, 1905)
Litoria infrafrenata (Günther, 1867)
Litoria modica (Tyler, 1968)
Litoria pygmaea (Meyer, 1874)
Litoria sp. 1
Litoria sp. 2
Litoria sp. 3
Nyctimystes cheesmani Tyler, 1964

Microhylidae
Callulops doriae Boulenger, 1888
Callulops slateri (Loveridge, 1955)
Callulops sp.
Cophixalus cheesmanae Parker, 1934
Cophixalus sp. 1
Cophixalus sp. 2
Copiula sp.
Hylophorbus rufescens Macleay, 1878
Mantophryne lateralis Boulenger, 1897
Oreophryne sp. 1
Oreophryne sp. 2
Oreophryne sp. 3
Sphenophryne cornuta Peters and Doria, 1878
Sphenophryne sp.
Xenobatrachus sp.

Reptiles recorded During the Survey

Chelidae
Emydura subglobosa (Krefft, 1876)

Trionychidae
Pelochelys bibroni (Owen, 1853)

Agamidae
Hypsilurus cf. auritus (Meyer, 1874)
Hypsilurus dilophus (Dumèril and Bibron, 1837)
Hypsilurus modestus (Meyer, 1874)

Gekkonidae
Cyrtodactylus cf. *mimikanus* (Boulenger, 1914)
Gehyra lampei Andersson, 1913
Gehyra membranacruralis King and Horner, 1989
Hemidactylus frenatus (Dumèril and Bibron, 1836)
Nactus sp.

Scincidae
Carlia fusca (Dumèril and Bibron, 1839)
Emoia caeruleocauda de Vis, 1892
Emoia kordoana (Meyer, 1874)
Emoia longicauda (Macleay, 1877)
Emoia pallidiceps pallidiceps de Vis, 1890
Emoia physicae purari Brown, 1991
Emoia physicina Brown and Parker, 1985
Emoia tropidolepis (Boulenger, 1914)
Eugongylus albofasciolatus (Günther, 1872)
Lipinia noctua (Lesson, 1830)
Lygisaurus novaeguineae (Meyer, 1874)
Prasinohaema cf. *flavipes* (Parker, 1936)
Prasinohaema prehensicauda (Loveridge, 1945)
Sphenomorphus jobiensis (Meyer, 1874)
Sphenomorphus leptofasciatus Greer and Parker, 1974
Sphenomorphus melanopogon (Dumèril and Bibron, 1839)
Sphenomorphus pratti (Boulenger, 1903)
Sphenomorphus muelleri (Schlegel, 1837)
Sphenomorphus stickeli (Loveridge, 1948)
Sphenomorphus solomonis (Boulenger, 1887)
Sphenomorphus sp.

Varanidae

Varanus indicus (Daudin, 1802)

Boidae

Candoia aspera (Günther, 1877)
Morelia albertisii (Peters and Doria, 1878)
Morelia viridis (Schlegel, 1872)

Colubridae

Boiga irregularis (Merrem, 1802)
Dendrelaphis calligastra (Günther, 1867)
Stegonotus cucullatus (Dumèril, Bibron and Dumèril, 1854)
Tropidonophis doriae (Boulenger, 1897)
Tropidonophis multiscutellatus (Brongersma, 1948)

Elapidae

Acanthophis antarcticus (Shaw and Nodder, 1802)
Aspidomorphus muelleri (Schlegel, 1837)
Micropechis ikaheka (Lesson, 1829)
Toxicocalamus sp. (near *loriae*)

Specimens are deposited in the PNG National Museum, Bishop Museum, U.S. National Museum and James Cook University.

Herpetofauna Species Accounts

These accounts are in the following format: general habitat, microhabitat, habits, food preferences, coloration, size (SVL), clutch or brood size, geographic distribution and notes. We have cited literature sources for specific information when this is not generally known.

Myobatrachidae

Lechriodus melanopyga (Doria, 1874)

Wet to dry forests; terrestrial; nocturnal; carnivorous, prey preferences unknown; head broad; dorsum extremely variable, with or without mid-dorsal stripe; SVL 50 mm; tadpole detritivore and possibly carnivore. Widespread in southern New Guinea; at Lakekamu a common inhabitant of the forest floor at night, and juveniles recorded from litter plots. Breeding occurred in shallow pools south of Ivimka Camp after heavy rains. Floating mats of frothy jelly were attached to vegetation, and these mats degenerated within several days releasing tadpoles into the pools. Mortality in small pools was 100% due to pool desiccation during this study. Tadpoles were collected from swamp 1 km south of Ivimka Camp for descriptive studies. Call "a series of loud snorts, reminiscent of the noise made by a pig" (Menzies 1973).

Ranidae

Rana arfaki (Meyer, 1874)

Forests; terrestrial, riparian, medium to large rivers; nocturnal; carnivorous, crabs, prawns, invertebrates (Menzies 1973); brown, males with warty skin, females with rough skin; toes fully webbed; largest New Guinea frog, heavy bodied with long powerful rear legs; SVL to 160 mm. Widespread throughout New Guinea, sea level up to about 1000 m. Breeding aseasonal, larval biology unknown, tadpole currently being described by R. A. Altig. At Lakekamu several specimens were observed at night, perched next to the Sapoi River adjacent to Ivimka Camp. One female was collected during mid-afternoon when it was disturbed on the river bank. Call "a double squeak uttered at rather long intervals" (Menzies 1973).

Rana daemeli (Steindachner, 1868)

Wet to dry forests; terrestrial; savanna and grasslands; riparian and still water, small to large streams and rivers, pools, swamps; carnivorous, prey preferences unknown; Sharp-snouted frog with long, powerful and distinctly banded rear legs; SVL to about 80 mm. Widespread in New Guinea and northern Queensland, Australia. Males call from banks adjacent to water or from emergent logs. Eggs laid in floating mass (Menzies 1987) and tadpole benthic (Richards 1992), probably detritivore. At Lakekamu a breeding chorus was heard almost every night in flooded swampy regrowth forest adjacent to Ivimka Camp. Call a loud series of three to six duck-like quacking notes.

Rana garritor (Menzies, 1987)

Forests; terrestrial and low vegetation; riparian, small streams; nocturnal; carnivorous, prey preferences unknown; Extremely sharp snout, long, powerful rear legs; morphologically similar to *Rana daemeli*; SVL to 80mm. Widespread in New Guinea to about 1400 m altitude. At Lakekamu several individuals called at night from leaves overhanging a small stream about 100 m from Ivimka Camp. Eggs attached in a single layer to stones in streams. Tadpoles not adequately described but have four tooth rows above and three below the mouth (Menzies 1973). The call is a rather loud series of harsh, rapidly repeated notes "crek crek crek" (Menzies 1973) lasting about 3-4 seconds.

Rana grisea (van Kampen, 1913)

Forests, open valleys, gardens; terrestrial; riparian, swamps; nocturnal; carnivorous, prey preferences unknown; sharp-snouted, long-legged frog with a distinct black mask across side of face; SVL to 90mm. Widespread in New Guinea from near sea level to over 1600 m. This "species" is in fact a complex of closely related taxa. Until mating calls are recorded at the type locality and the status of the various populations is addressed, the Lakekamu population is only tentatively assigned to this species. At Lakekamu juveniles were abundant in litter plots and VES surveys. Adults were observed adjacent to the Sapoi River at night, and one adult was observed basking in a sun patch in mid-morning. The call of the Lakekamu species was not recorded.

Hylidae

Litoria dorsalis (Macleay, 1878)

Wet and open forests; scansorial to arboreal; swamps; nocturnal; prey preferences unknown; small, slender species, SVL to about 20 mm. Southern, lowland PNG, southern Fly River region. Larval biology unknown. At Lakekamu associated with swamps in primary forest. Abundant at swamp 1 km south of Ivimka Camp where it formed choruses with *L. pygmaea* and *Litoria* sp. near *timida*. Call a short, high-pitched chirp repeated at irregular intervals.

Litoria genimaculata (Horst, 1883)

Forests; arboreal; riparian and swamps; nocturnal; carnivore, wide range of invertebrates. Mottled brown and green, distinct crenulated ridge along outside of arm and leg; SVL to about 50 mm. Widespread in lowlands and hills of New Guinea and northern Queensland to about 1500 m altitude. Status of Australian populations currently under investigation. At Lakekamu associated with both riparian and swamp habitats. Call a series of soft notes "toc toc toc..."

Litoria graminea (Boulenger, 1905)

Forests; arboreal, canopy dweller; nocturnal; carnivore, prey preferences unknown. Bright green frog with large toe pads and extensively webbed fingers, SVL 65 mm. Widespread in lowlands of New Guinea up to ca. 300 m elevation. Biology unknown due to inaccessible microhabitat. At Lakekamu males were calling most nights during the survey, most from over 20m above ground in primary forest. One calling individual was found 5m above the ground. Call a single, distinctly pulsed note "craak" repeated at approximately 3 second intervals.

Litoria infrafrenata (Günther, 1867)

Forests, savanna, gardens, human habitation; riparian, swamps; arboreal; nocturnal; carnivore, prey preferences unknown. Large, slender green to brown frog with distinct white stripe on lower lip. SVL to 135 mm. Widespread in lowland New Guinea, eastern Indonesia and northern Queensland, Australia. At Lakekamu several individuals were found in association with small streams and swamps. Call a loud double note repeated regularly for long periods.

Litoria modica (Tyler, 1968)

Forest; riparian, swift streams; arboreal; nocturnal; carnivore, prey preferences unknown; small frog, SVL to about 35 mm. Tadpoles with large, ventral suctorial mouth, currently being described by R. A. Altig. This "species" is a complex of morphologically similar taxa that are widespread across foothills and mountains of New Guinea. The status of the Lakekamu populations requires verification. Call of the Lakekamu species not recorded.

Litoria pygmaea (Meyer, 1874)

Forest; swamps; arboreal; nocturnal; carnivore, prey preferences unknown. SVL to 45 mm. Known from widely scattered localities in western and southern New Guinea. Lakekamu is a significant easterly range extension. At Lakekamu males formed dense choruses at the swamp 1 km south of Ivimka Camp. Males called from 1-2 m above the ground, on leafy branches adjacent to or overhanging the swamp. Call a loud, high pitched bleating sound in which each note appears to have about 3-4 distinct pulses.

Litoria sp. 1

Forest; riparian; arboreal; nocturnal; carnivore, prey preferences unknown. SVL to 20 mm. Green with yellow stripe on lip. An undescribed species of uncertain affinities. Morphologically similar to members of the *Litoria bicolor* group but males called from perch sites 2-4 m over rapidly flowing sections of the Sapoi River upstream from Ivimka Camp, indicating relationships may be closer to torrent frogs of the *Litoria modica* group. Larval biology unknown. Call with two distinct components: a short rasping note, uttered singly or in short series; a series of spluttering notes terminate the call sequence.

Litoria sp. 2

Forest; riparian, swamps, small streams; arboreal; nocturnal; carnivore, prey preferences unknown; SVL to 35 mm; bright green dorsum with numerous dark green spots. An undescribed species closely related to *Litoria gracilenta*. At Lakekamu found at two sites along a tributary of the Avi Avi, where it formed aggregations around stream pools. At Ok Tedi, Western Province, this species occurs adjacent to swamps. Call of Lakekamu population not recorded.

Litoria sp. 3

Forest; swamp; arboreal; nocturnal; carnivore, prey preferences unknown. SVL to 25 mm. Probably an undescribed species, but closely related to *Litoria timida*. At Lakekamu this species formed choruses with *L. pygmaea* and *L. dorsalis* at the swamp 1 km south of Ivimka Camp. Males called from elevated positions on vegetation around the swamp perimeter. Call a series of mechanical rasping buzzes.

Nyctimystes cheesmani (Tyler 1964)

Forest; riparian, small to large, swift streams; arboreal; nocturnal; carnivore, prey preferences unknown. SVL to 50 mm. Pale brown frog with lichen-shaped dorsal markings and extremely large eyes with vertical pupil and reticulated lower eyelid. *N. cheesmani* is a complex of morphologically similar species, and this taxon is assigned only tentatively to this species. At Lakekamu males were collected at night from low vegetation overhanging small and large creeks. A calling male was observed perched on a large rock in the center of the Sapoi River upstream from Ivimka Camp. A tadpole with suctorial mouthparts collected from the Sapoi River almost certainly represents this species. Call of the Lakekamu population not recorded.

Microhylidae

Callulops doriae (Boulenger, 1888)

Forest; terrestrial; nocturnal; carnivore, prey preferences unknown. SVL to 100 mm. Robust, short-legged frog. Brown with distinct darker warts dorsally, many with white centers. Widespread in eastern PNG, from sea level to over 1800 m. At Lakekamu males called from semi-concealed positions among tree buttresses. One calling male was covered with mosquitoes. Releases copious amounts of sticky white exudate when disturbed. Call a series of 2-3 deep, rather resonant barks that can be heard for up to 800m.

Callulops slateri (Loveridge, 1955)

Forest; arboreal; nocturnal; carnivore, prey preferences unknown. SVL to 55 mm. Dorsally brown with irregular darker markings and often with thin, pale middorsal stripe. Known from widely scattered localities in the lowlands and foothills of southern PNG. Calls from elevated perch sites in *Pandanus* trees and tree ferns, the leaf axils of which appear to be oviposition sites. A female was observed to initiate mating by hopping onto a male frog. This elicited a different (abbreviated, softer) courtship call by the male, which hopped approximately 15 cm away and resumed calling until the female reinitiated contact. Call a short, distinctly pulsed musical trill. This species was not previously known east of the Kikori Basin. Our specimens therefore extend the known range of this species more than 200 km eastwards.

Callulops sp.

Forest; terrestrial, burrows; nocturnal; carnivore, prey preferences unknown. SVL to at least 43 mm. Uniform brown, short-legged and robust frog. An undescribed species. At Lakekamu occurred in high densities around Ivimka Camp, especially in disturbed vegetation around camp. Males called from slightly elevated positions at the base of trees or shrubs. Call a short series of about 6-10 harsh barking notes.

Cophixalus cheesmanae (Parker, 1934)

Forest; scansorial; nocturnal; carnivore, prey preferences unknown. SVL to 30 mm. A slender, light brown and sharp-snouted frog. Known from widely scattered localities in southern Irian Jaya and eastern PNG. At Lakekamu was found exclusively in riparian habitats, especially along small streams. Males and females were observed at night perched on vegetation 1-3 m above the ground. Call a "rather quiet, slow stuttering noise without any musical quality" (Menzies 1973).

Cophixalus sp. 1

Forest; arboreal, found mostly in moss and epiphytic growth on tree trunks; predator, prey preferences unknown, presumably small insects; mostly brown; SVL 20 mm; reproductive habits unknown. This is an undescribed species of uncertain affinities.

Cophixalus sp. 2

Forest; arboreal. SVL 15 mm. An undescribed species, one specimen collected from within a *Myrmecodia* plant 5 m above the ground. Call not recorded.

Copiula sp.

Forest; terrestrial, litter; nocturnal; carnivore, prey preferences unknown. SVL 29 mm. Possibly an undescribed species that was common in litter throughout the Ivimka Camp area. The status of this species requires confirmation.

Hylophorbus rufescens (Macleay, 1878)

Forest; terrestrial, litter; nocturnal; carnivore, prey preferences unknown. SVL to 45 mm. Slender, brown frog with black markings laterally. Widespread throughout New Guinea and offshore islands from sea level to over 3500 m. Common at Lakekamu, one of the most abundant species in the litter plots. *H. rufescens* is a complex of morphologically similar species, and this taxon is included here only tentatively. Call a continuous series of soft, slightly musical notes "arp arp arp" repeated at a rate of approximately one per second.

Mantophryne lateralis (**Boulenger, 1897**)

Forest; terrestrial, litter; nocturnal; carnivore, prey preferences unknown. SVL to 55 mm. Brown with broad light mid-dorsal band and dark lateral bands. Widespread throughout lowlands of New Guinea up to about 1200 m altitude. At Lakekamu several individuals were found calling from exposed positions on the surface of forest litter. Call a long series of harsh barking notes.

Oreophryne **sp. 1**

Forest; scansorial to arboreal; nocturnal; SVL 25 mm. Undescribed species known elsewhere from Ok Tedi headwaters and Crater Mountain. Johnston and Richards (1993) report that this species lays eggs attached to leaves 2-4 m above the ground, which are guarded by the male. Common around Ivimka Camp. Call a short, loud rattle repeated at irregular intervals.

Oreophryne **sp. 2**

Forest; arboreal; nocturnal; SVL 25 mm. Probably undescribed species in *O. biroi* group. At Lakekamu heard calling from over 15 m above ground in one patch of forest. A single specimen, presumably representing the calling individuals, was collected on a low shrub.
Call a series of about 10-15 musical peeps.

Oreophryne **sp. 3**

Forest; arboreal; nocturnal; SVL to at least 23 mm. Probably an undescribed species related to *O. loriae*. An abundant species at Lakekamu where males called from dense foliage more than 5 m above the ground. Call a short rasping buzz repeated about 6-12 times.

Sphenophryne cornuta (**Peters and Doria, 1878**)

Forest; scansorial; nocturnal. SVL to 35 mm. Snout pointed, sharp spine-like tubercle on each eyelid. Widespread throughout lowland New Guinea, to an altitude of about 1200 m. An uncommon frog at Lakekamu where several males were found calling from low bushes up to about 2.5 m above ground. Call a loud, explosive rattle that could be confused with *Oreophryne* sp. 1.

Sphenophryne **sp.**

Forest; terrestrial, litter; nocturnal, also calling at dusk. SVL 45 mm. Olive green with small, raised black warts. An undescribed species, the Lakekamu population is the easternmost known locality for this species, which is known from as far west as the Purari drainage. It is currently being described by R. G. Zweifel. Call an extremely long series of rapidly repeated yapping notes.

Xenobatrachus **sp.**

Forest; fossorial, beneath litter; nocturnal. SVL 33 mm. A species of uncertain affinities, possibly undescribed. Like other members of this genus males call from beneath the soil/litter layer. At Lakekamu several specimens were collected in litter plots. One calling male was collected at night from beneath litter at the base of a worm mound. Two individuals, one of which was a prey item of a *Ptilorrhoa caerulescens* captured in a mist net, possibly represent a second species. Call a series of extremely soft, melodious notes "hoot..... hoot...."

Chelidae

Emydura subglobosa (Krefft, 1876)

Prefers rivers but is also found in lakes and lagoons; aquatic, occasionally basking; carnivorous; crustaceans, aquatic insects and molluscs; carapace broadly oblong, slightly wider posteriorly, dull brown, plastron mostly yellow and edged in red, undersides of marginals also red; carapace length to 250 mm; oviparous; clutch size ca. 10 eggs; coastal regions of mainland PNG, southern regions of Irian Jaya, and northern regions of Australia (Georges and Adams, 1996). This species is abundant in the lower reaches of the Lakekamu Basin in slow meandering rivers. We found only a small juvenile around the Ivimka camp, suggesting that *E. subglobosa* is rare in that area.

Trionychidae

Pelochelys bibroni (Owen, 1853)

Slow meandering rivers and streams, mostly in deep water, occasionally in brackish water; highly aquatic; mostly carnivorous; crustaceans, molluscs; carapace uniform dull olive-brown, plastron cream-colored; carapace length to ca. 1100 mm; oviparous; clutch size ca. 25 eggs; originally thought to occur throughout New Guinea and much of southeast Asia, but as a result of a recent taxonomic redescription this species is now endemic to southern New Guinea. All other populations of *Pelochelys,* including those in northern New Guinea are tentatively referred to *P. cantorii* Gray, 1864 (Rhodin *et al.* 1993; Webb, 1995). *Pelochelys* are the largest softshell turtles in the world and the largest freshwater turtle in New Guinea. A single individual estimated to a have a mass of at least 25 kg was observed in a deep pool in the Avi Avi River 4.6 km south of Ivimka camp.

Agamidae

Hypsilurus cf. *auritus* (Meyer, 1874)

Forests, almost exclusively in riparian microhabitats; highly arboreal; foraging habits unknown; prey preferences also unknown; dorsum brown to dark brown with black reticulations, black spot on each side of the head, tail and limbs often with black banding, venter brown to yellowish brown; SVL 125-130 mm; clutch size unknown; recorded from widely scattered localities on the north and south coasts of New Guinea, mostly in Irian Jaya. The systematics of *Hypsilurus* are poorly understood and we tentatively refer our material to *H. auritus*. This requires confirmation as it would represent a considerable range extension of the species along the south coast of New Guinea. This species was almost always observed in close association with streams, often on overhanging branches, and would dive into the water when pursued.

Hypsilurus dilophus (Dumèril and Bibron, 1837)

Forests and regrowth areas; arboreal; foraging habits not known; prey preferences also unknown but probably including insects and small fruits; dorsum dull reddish-brown with faint black transverse bands, venter light brown, inside of mouth bright orange; SVL 220-230 mm; oviparous, clutch size unknown; occurs throughout New Guinea and parts of Maluku (=Moluccas) from sea level to ca. 800 m elevation.

Hypsilurus modestus (Meyer, 1874)

Forests; arboreal; moves slowly through the tree canopy; insects and possibly small fruits; dorsum generally bright green, occasionally darkening to greenish-brown, venter off-white to cream; SVL 90-100 mm; oviparous; clutch size 2 to at least 5; Admiralty and Bismarck Archipelago and most of lowland New Guinea, also the Aru Islands at elevations below 500 m. This species is probably common in the Lakekamu Basin but is infrequently observed because of its arboreal habits and cryptic coloration.

Gekkonidae

Cyrtodactylus cf. *mimikanus* (Boulenger, 1914)

Primary and secondary forest; arboreal, nocturnal; sit and wait predator; insects and possibly small lizards; the dominant color pattern in our specimens was dorsum dull yellowish-brown with broad dark brown transverse bands, these often edged with white, venter gray-white with yellowish tinge; SVL 90-100 mm; oviparous; clutch size probably 2; endemic to New Guinea where it has been recorded from scattered low-land localities on both the north and south coasts Because of its arboreal habits, this species is infrequently observed throughout what is probably an extensive range. Charles Burg (personal communication) saw an adult during the day on a tree trunk 2-3 m above the ground along the track from Tekadu to Ivimka. McCoy (1980) reports that a close relative, *C. louisiadensis*, prefers large trees, particularly *Ficus*. Although our material keys to *C. mimikanus* in Bauer and Henle (1994) on the basis of scale counts, the color patterns don't match *C. mimikanus* and it is likely that our specimens represent an undescribed species.

Gehyra lampei (Andersson, 1913)

Disturbed areas; arboreal, nocturnal; sit and wait predator; dorsum gray-brown with slight olive tinge (sometimes it lightens considerably under the control of the animal), a series of transverse bands begin-ning in neck region and continuing along tail - these initially (i.e., neck region) composed of a series of light yellowish blotches, irregularly edged in black - extending to flanks, on the tail these bands are mostly black, there is also a pale thin yellowish vertebral line, venter bright lemon yellow - duller on tail, iris golden; SVL 60 mm; presumably oviparous with a clutch size of two; known previously only from the type locality, Bogadjim on the north coast of New Guinea (Bauer and Henle, 1994).

Gehyra membranacruralis (King and Horner, 1989)

Forests and second growth areas; arboreal, nocturnal; sit and wait predator; insects and possibly small lizards; dorsum uniform sandy-tan with two conspicuous yellow spots in shoulder area, venter mostly yel-low, whiter in distal areas; SVL 130+ mm; oviparous, clutch size two; currently known with certainty from the Port Moresby area. This species requires confirmation. King and Horner (1989) suggest that most specimens of the *Gehyra vorax* recorded from New Guinea may actually be this species. Beckon (1992) on the other hand included *G. membranacruralis* within his New Guinea morphotype of *Gehyra vorax*. Bauer and Henle (1994) tentatively recognizes *G. membranacruralis* as valid, and points out that resolu-tion of this confusing situation will require careful study of type material.

Hemidactylus frenatus (Dumèril and Bibron, 1836)

Disturbed areas, including village houses; arboreal, nocturnal; sit and wait predator; insects; dorsum pale cream-white to gray brown, sometimes with brown blotches, this varying with temperature and physiolog-ical state, venter white to gray-white; SVL 50-60 mm; oviparous, clutch size 2; native to southeast Asia. This species is a human commensal that has been transported to settlements throughout the tropical world. It is abundant around Tekadu.

Nactus sp.

Forests; arboreal/terrestrial, nocturnal, often seen perched on the base of large trees terrestrial; sit and wait predator; insects; dorsum gray to gray-brown with wavy dark brown to black bands, venter gray-white; SVL 60-65 mm; oviparous; clutch size probably 2; geographic and elevational range unknown. This species would key to *Nactus pelagicus* (Girard 1858) in Bauer and Henle (1994), but that taxon is now thought to comprise a superspecies with many unisexual and bisexual forms distributed throughout much of the western Pacific and northeastern Australia (Zug and Moon 1995). Our collections included males indicating that the Lakekamu population represents a bisexual species.

Scincidae

Carlia fusca (Dumèril and Bibron, 1839)
Forest interior and edge, mostly in open, sunny areas; terrestrial; active forager; insects; dorsum overall drab olive-brown with white and black flecks, venter gray-white; 45-50 mm SVL; oviparous; clutch size 2-3; western Solomon Islands east throughout New Guinea north to Palau and the Northern Marianas and the Bismarck and Admiralty islands at elevations from sea level to at least 1100 m. This species is abundant in the Lakekamu Basin and is probably the most common lizard in open, sunny areas. *Carlia fusca* is probably a large superspecies complex and is currently under revision by G. Zug; the taxon inhabiting the Lakekamu Basin would key to *Carlia luctuosa* (Peters and Doria, 1878) in Loveridge (1948).

Emoia caeruleocauda (de Vis, 1892)
Forest interior and edge; terrestrial but sometimes ascends 1-2 m into underbrush and tree buttresses; active forager; insects; dorsum dark brown-black with five thin white to gray-white longitudinal stripes from the head region to the base of the tail, tail is often bright blue, venter whitish with a pearl sheen; 40-60 mm SVL; oviparous; clutch size 2; throughout New Guinea and much of the southwest Pacific Basin north to Micronesia and the southern Philippines and east to Fiji at elevations mostly below 1500 m; this species is abundant in the Lakekamu Basin and is often seen basking in patches of sunlight on the forest floor. It is also common in disturbed areas such as village clearings.

Emoia kordoana (Meyer, 1874)
Forest interior and edge; arboreal; active forager; insects; dorsum overall dull olive-brown with white and black flecks, venter light yellow; 45-60 mm SVL; oviparous; clutch size 2; throughout New Guinea, Aru, Bismarck and Admiralty islands at elevations mostly below 500 m; this species is common in the Lakekamu Basin but is not often seen because of its arboreal habits; most commonly seen in disturbed areas at the forest edge.

Emoia longicauda (Macleay, 1877)
Forest interior and edge; arboreal; active forager; insects; dorsum dull tan-brown, mostly uniform but sometimes with a few black flecks, venter light yellowish-green; 65-100 mm SVL; oviparous; clutch size 2; occurs throughout New Guinea and offshore islands, including Fergusson, Woodlark and the Louisades, also Admiralty Archipelago as well as the islands of the Torres Strait and the Cape York Peninsula of Australia at elevations mostly below 1500 m. This species is similar to *Emoia cyanogaster* which has similar habits but is allopatric, occurring throughout most of the Bismarck archipelago east to the Solomon Islands and Vanuatu. *Emoia longicauda* is common in the Lakekamu Basin where it is most frequently seen in small trees at the forest edge in garden clearings, *etc*. It is larger and more heavy-bodied than *E. kordoana* which occurs in the same habitats.

Emoia pallidiceps pallidiceps (de Vis, 1890)
Forest interior and edge, village clearings and associated disturbed areas; terrestrial; active forager; insects; dorsum tan - olive-tan with paravertebral broken band of brownish spots; upper flanks bordered at the bottom by a white longitudinal, continuous or mostly continuous band extending from the ear to the hind legs, lower flanks light gray-tan, venter similar but lighter; SVL 40-50 mm; oviparous, clutch size 2; endemic to PNG and recorded from Gulf Province east to Milne Bay Province and along the north coast to Madang Province and throughout most of the highlands provinces where it occurs from sea level to at least 1800 m elevation. We collected only one specimen from the Lakekamu Basin and assume the species to be rare in this area. However, this represents a slight westward extension of the known range of the species on the south coast of New Guinea.

Emoia physicae purari (Brown, 1991)

Forest interior and edge, garden clearings and other disturbed areas; mostly terrestrial (occasionally ascends 1-2 m off the ground on tree buttresses); active forager; insects; dorsum, including flanks and tail, mostly uniform tan-brown with scattered flecks of dark brown-black and whitish flecks, venter light gray; SVL 55-75 mm; oviparous, clutch size 2; endemic to PNG where it was previously known only from the Purari River drainage at elevations from sea level to ca. 1200 m. Our collections extend the range of the species ca. 200 km eastward.

Emoia physicina (Brown and Parker, 1985)

Forest interior; terrestrial; active forager; insects; dorsum, mostly uniform brown with scattered flecks of dark brown in paravertebral region, and scattered yellowish flecks overall, upper flanks darker brown, lower flanks gray-brown, venter silver; SVL 40-50mm; oviparous, clutch size 2; endemic to New Guinea where it occurs from the Mimika River in Irian Jaya east to the Central Highlands of PNG and in the lowlands of Western and Gulf provinces from sea level to ca. 1200 m. Our collections extend the range of the species ca. 200 km eastward.

Emoia tropidolepis (Boulenger, 1914)

Forest interior and edge, village clearings and associated disturbed areas; terrestrial; active forager; insects; dorsum mostly uniform light brown with slight hint of transverse banding, venter mostly gray-white; SVL 50-70 mm; oviparous, clutch size 2; endemic the south coast of New Guinea where it was previously known from the Mimika River drainage in Irian Jaya east to the Kikori River drainage in PNG at elevations mostly below 500 m. We collected only one specimen, a juvenile, which we tentatively assign to this species. This extends the known range ca. 250 km eastward. We assume the species to be rare in the Lakekamu Basin.

Eugongylus albofasciolatus (Günther, 1872)

Forest and regrowth areas with shade; terrestrial, crepuscular; insects and other lizards; dorsum light to dark yellowish-brown with faint light yellowish-brown transverse bands, lateral margins of dorsal scales edged in black giving appearance of thin black longitudinal stripes, flanks similar to dorsum, prominent black "slashes" extending from upper labials to chin and throat, venter yellowish-cream; SVL 150-160 mm; oviparous; clutch size up to 5; occurs from the Solomon Islands east to New Guinea and the Cape York Peninsula of Australia to Maluku (=Moluccas). This species appears to be uncommon throughout its range.

Lipinia noctua (Lesson, 1830)

Open, disturbed areas; arboreal, sit and wait predator; insects; dorsum dark golden brown with black flecks, dark yellowish spot in neck region and extending posteriorly as a broad vertebral band edged with black to dark brown, venter yellowish-white; SVL 40-50 mm; probably viviparous; litter size probably two; this species is known throughout New Guinea, Maluku (= Moluccas) and much of the Pacific Basin. However, *L. noctua* is probably a superspecies complex and it is not certain that our specimens are referable to the nominate form.

Lygisaurus novaeguineae (Meyer, 1874)

Forest interior; terrestrial, diurnal and crepuscular, active forager; insects; dorsum uniform brown with yellowish, indistinct dorsolateral line, flanks dark brown, lips white with black spots, venter, including chin and throat, white; SVL 30-35 mm; oviparous; clutch size 2; widespread throughout the south coast lowlands of New Guinea and Maluku (=Moluccas) at elevations below 500 m. Ingram and Covacevich (1988) treat the taxon previously known as *L. novaeguineae* from the Torres Strait as *L. macfarlani* (Günther, 1877). They argue that *L. novaeguineae* cannot be "applied convincingly to any of the

Lygisaurus known from New Guinea". We use this name simply because it is in widespread use for the *Lygisaurus* inhabiting the south coast of New Guinea. The Lakekamu taxon is probably the most common lizard in the basin.

Prasinohaema cf. *flavipes* (Parker, 1936)

Primary and secondary forest; arboreal; active, slow-moving predator; insects; coloration of P. flavipes is highly variable, in our single specimen dorsum olive-brown with indistinct, broad, dark brown transverse bands, these often edged in light brown and extending to lower flanks; side of face to front leg yellowish-green, dark-brown bands very prominent - broad dark brown longitudinal band extending from loreal region through eye to just above ear opening, tissue in mouth and throat pale green, flanks similar to dorsum but generally noticeably lighter brown with a tinge of gray, venter off-white to gray with a tinge of green, tail similar to rest of venter; SVL 105 mm; ovoviviparous; litter size unknown; *Prasinohaema flavipes* is known mostly from montane regions throughout New Guinea at elevations of 1200 m and above. Our specimen was collected at ca. 400 m elevation (Tekadu) and is undoubtedly distinct from *P. flavipes*.

Prasinohaema prehensicauda (Loveridge, 1945)

Primary and secondary forest; arboreal; active, slow-moving predator; insects; coloration variable, in our single specimen, a female, dorsum tan-brown, mottled with white, especially in neck region and side of face, several prominent brown slashes from back of jaw to just posterior of the tympanum, eyelids white, three prominent evenly spaced white spots along dorso-lateral area, flanks similar in coloration to dorsum, venter pale white with yellowish tinge, males of this species are mostly green (Woodruff 1972); SVL 50-55 mm; ovoviviparous; litter size unknown; *Prasinohaema prehensicauda* is known mostly from montane regions in the Central Highlands of New Guinea at elevations of 1500 m and above. Our specimen was collected around ca. 400 m elevation (Tekadu) and would represent a considerable lowland range extension for this species; possibly our specimen represents a closely related taxon.

Sphenomorphus jobiensis (Meyer, 1874)

Forest interior; terrestrial, crepuscular, active predator; insects; dorsum brown with faint transverse black bars - these showing a slight tendency to develop into "chevrons", head uniform brown, similar in coloration to rest of body, large black blotch above ear, this extending forward in an irregular, narrow line to eye, dorso-lateral region with a series of tan blotches "bracketed" in black, upper flanks gray-brown with faint longitudinal black streaking and with distinctive salmon tinge from side of head to mid-body, venter, including chin and throat, yellowish, tail white-pearl from vent; SVL 80-85 mm; oviparous; clutch up to 4 eggs; throughout lowland New Guinea to ca. 1200 m. It is likely that this taxon is a superspecies (Donnellan and Aplin 1989).

Sphenomorphus leptofasciatus (Greer and Parker, 1974)

Forest edge and second growth areas; terrestrial; secretive, probably crepuscular, predator; insects; dorsum light to dark brown, with numerous thin yellowish crossbands, these particularly distinctive in the young, flanks similar to dorsum, venter brown - gray-brown; SVL 60-80 mm; ovoviviparous, litter size 1-6; found throughout the Central Highlands of PNG but recorded also from Aseki and adjacent areas of the south coast at elevations from ca. 1000-1800 m. The material from Lakekamu represents a significant lowland range extension for this species.

Sphenomorphus melanopogon **(Dumèril and Bibron, 1839)**

Forest interior; terrestrial, crepuscular, active predator; insects and small lizards; dorsum brown with faint transverse black bars - these showing a slight tendency to develop into "chevrons", head brown, darker than rest of body, dorso-lateral region lighter, but not particularly distinct; upper flanks similar to dorsum, faintly suffused with lighter blotches, with slight tendency of transverse bands to form black blotches in dorso-lateral area, lower flanks similar to upper flanks, but distinctly lighter overall, chin and throat salmon with fine back longitudinal streaking - this extending to pectoral region, rest of venter, including tail, light cream; SVL 95-105 mm; oviparous; clutch size 2-4; *Sphenomorphus melanopogon* is known from New Guinea and associated offshore islands west to Sulawesi. However, the type series probably represents more than one species and extensive taxonomic study will be required to determine the correct name that applies to the New Guinea populations.

Sphenomorphus pratti **(Boulenger, 1903)**

Forest edge and disturbed areas; terrestrial, fossorial, diurnal (?); insects; dorsum mottled olive-brown/pale olive brown, head and neck more or less uniform olive brown, tail mottled with gray, side of face with alternating light and dark slashes, venter cream except for chin which has gray striations; SVL 90-100 mm; oviparous, clutch size 3-5; known mostly from mid-montane areas (1000-1400 m) in eastern New Guinea.

Sphenomorphus muelleri **(Schlegel, 1837)**

Forest interior; terrestrial, diurnal and crepuscular, active predator; insects and small vertebrates; dorsum brown with black speckling, dark brown-black dorsolateral band edged in yellow from snout to groin, flanks and venter yellowish, throat pale brown; SVL 150-180 mm; reproductive habits unknown, probably oviparous; clutch size unknown; widespread throughout much of New Guinea and Maluku (=Moluccas) but infrequently seen.

Sphenomorphus stickeli **(Loveridge, 1948)**

Forest interior and edge; terrestrial, diurnal and crepuscular, sit and wait predator; insects; dorsum golden-brown with irregular cream-colored crossbands in lateral areas and an irregular series of black specks and spots in paravertebral region, thick chocolate brown dorso-lateral band, venter, including chin and throat, cream-white; SVL 45-50 mm; oviparous; clutch size 2; Bismarck Archipelago and New Guinea, where it occurs throughout much of the north lowlands and in scattered localities along the south coast at elevations from sea level to 670 m. It occurs to at least 1000 m elevation in New Ireland (Mys 1988).

Sphenomorphus solomonis **(Boulenger, 1887)**

Forest interior; terrestrial, fossorial, crepuscular or nocturnal (?), secretive predator; insects; dorsum reddish brown, suffused with black specks; tail and upper flanks similar, slightly darker; lower flanks similar but distinctly lighter; venter dark reddish-orange suffused with gray; SVL 50 mm; oviparous, clutch size 3; known from Maluku (Halmahera, Ternate and Morotai) east across the north coast of New Guinea to most of the islands in the Solomons Archipelago at elevations from sea level to 1650 m (Mys 1988). We tentatively refer our material to *S. solomonis*. Although this species has apparently been recorded from Gulf Province (Whitaker and Whitaker 1982), its occurrence south of the main ranges is unexpected, and it is possible that our material represents a closely related taxon.

Sphenomorphus sp. (*jobiensis* group)
Forest interior; terrestrial, crepuscular, active predator; insects and small lizards; dorsum brown with faint transverse black bars - these showing a slight tendency to develop into "chevrons", head reddish-brown, side of face with large irregular black blotches, these forming vertical bars on side of neck, rest of dorso-lateral region/upper flanks with a series of large black blotches coalescing into a longitudinal series, this bordered on the top by light brown, forming something of a discontinuous dorso-lateral line, lower flanks with an irregular series of transverse black bands - oriented backwards (bottom end anterior to top end), chin and throat white with a single black line along the lateral edges and some irregular black blotches medially, rest of venter reddish - salmon; underside of tail lighter; SVL 100 mm; probably oviparous; clutch size probably 2-4; distribution unknown.

Varanidae

Varanus indicus (Daudin, 1802)
Forest and regrowth areas; terrestrial and arboreal; active, fast-moving predator; small vertebrates, including birds and their eggs; dull olive-green dorsally suffused with yellow specks, venter cream to yellow; SVL to ca. 500 mm; oviparous; clutch at least ca. 5 eggs, probably increasing with female body size; known from the Solomon Islands east to New Guinea and associated island groups to Maluku (= Moluccas) and north to Palau, the Northern Marianas and a number of Micronesian islands in the Caroline and Marshall groups. This species is very similar to *V. jobiensis* which also occurs throughout much of New Guinea.

Boidae

Candoia aspera (Günther, 1877)
Wet forest and regrowth areas; terrestrial, nocturnal; sit and wait predator; small vertebrates; heavy bod-ied, dorsum dark brown to reddish brown, with light brown to orange irregular banding, venter mostly pale reddish brown to black; SVL to ca. 600-800 mm; ovoviparous; litter size 12-18; known from the Admiralty and Bismarck Archipelagos and most of New Guinea. The nearest reported south coast locality from the Lakekamu Basin is in the Berina area near Port Moresby; our sighting therefore extends the range of *C. aspera* about 100 kilometers westwards along the south coast of PNG. See range map in O'Shea (1996: 202).

Morelia albertisii (Peters and Doria, 1878)
Wet forest and regrowth areas; terrestrial, nocturnal; sit and wait predator; small mammals; dorsum mostly glossy light brown to black with iridescent sheen , lighter on flanks, venter white to cream; SVL to ca. 1000-1800 mm; oviparous; clutch size 8-15; known from throughout New Guinea and offshore islands including those of the Torres Strait.

Morelia viridis (Schlegel, 1872)
Wet forest and regrowth areas; arboreal, occasionally terrestrial, nocturnal; sit and wait predator; small vertebrates; bright emerald green with broken series of yellow spots along the vertebral line and scattered white flecks on the flanks, young rust-red, changing to yellow and then green as they mature; SVL to ca. 1100 mm; oviparous; clutch size 13-26 eggs; known from throughout New Guinea and rain forests of the Cape York Peninsula, Australia.

Colubridae

Boiga irregularis (Merrem, 1802)

Primary and secondary forest, including regrowth areas and buildings; mostly arboreal; occasionally terrestrial, nocturnal and crepuscular; active predator; small vertebrates; dorsum mostly yellowish brown to reddish brown with faint transverse black bands, venter uniform reddish brown, iris yellow with vertical black pupil; SVL to ca. 1500-1800 mm; oviparous; clutch size 6-8 eggs; occurs from the Solomon Islands east throughout New Guinea to Maluku (=Moluccas) and northeastern Australia at elevations from sea level to 1400 m. This species is rear-fanged and mildly venomous.

Dendrelaphis calligastra (Günther, 1867)

Primary and secondary forest, including regrowth areas; arboreal and terrestrial, diurnal; active predator; small vertebrates; coloration highly variable, dorsum mostly blue-green to brown, venter cream to pale brown, generally with a broad black band from the snout through the eye to anterior flanks; SVL ca. 1100 mm; oviparous; clutch size 4 eggs; occurs through much of New Guinea and associated islands as well as northeastern Australia from sea level to ca. 1200 m. This snake is very fast moving and difficult to capture.

Stegonotus cucullatus (Dumèril, Bibron and Dumèril, 1854)

Primary and secondary forest, including regrowth areas; mostly terrestrial, nocturnal and crepuscular; active predator; small vertebrates including eggs of lizards and snakes; coloration variable, dorsum mostly uniform slate-gray but in some populations scales edged in black, venter pale gray; SVL ca. 1000 mm; oviparous; clutch size of 7-8; occurs throughout much of PNG, including Bougainville, to Irian Jaya and northeastern Australia.

Tropidonophis doriae (Boulenger, 1897)

Primary and secondary forest, mostly in riparian habitat; mostly aquatic, mostly nocturnal; active predator; mostly frogs, also fish and tadpoles; coloration variable, dorsum tan to bright orange with broad black transverse bands, venter similar, paler and without black bands, coloration of one of our specimens was dorsum light brown with alternating black bands - in the mid-body region these bands occasionally not meeting in the middle, head dark brown, tail also mostly dark-brown-black, venter cream, uniform, underside of tail similar but with diffuse black specks laterally; SVL ca. 800-1000 mm; oviparous; clutch size 3-8; Aru Islands and throughout much of New Guinea (except for the Vogelkop). Commonly seen at night in ponds and slow-moving streams and rivers in the Lakekamu Basin.

Tropidonophis multiscutellatus (Brongersma, 1948)

Primary and secondary forest, mostly in riparian habitat; mostly aquatic, diurnal and nocturnal; active predator; mostly frogs, also fish; coloration variable, dorsum uniform brown to gray, sometimes with irregular black spots, venter similar, much paler; SVL ca. 800-950mm; oviparous; clutch size 2-7; throughout much of New Guinea from sea level to ca. 1400 m. The only individual seen and collected was active in mid-morning on the forest floor adjacent to the Bulldog Road 1.8 km south of Ivimka Camp. It attempted to hide under leaf litter when pursued.

Elapidae

Acanthophis antarcticus (Shaw and Nodder, 1802)

Wet forest, mostly shaded areas; terrestrial; sit and wait predator; small vertebrates, particularly lizards; coloration extremely variable, dorsum generally brown to gray with transverse bands of similar but somewhat lighter color, these often including black specks and spots, head similar to rest of body, labials mostly black irregularly edged in white, venter generally white, often suffused with black spots; SVL 400 - 600 mm; ovoviparous; litter size up to 20; known from throughout New Guinea and Australia. The systematics of *Acanthophis* in New Guinea are poorly understood. Cogger (1992) lists *A. antarcticus* as an endemic Australian species and a close relative, *A. prelongus* Ramsay, 1877, as occurring in northern Australia and New Guinea. O'Shea (1990, 1996) mentions both *A. antarcticus* and *A. prelongus*, discusses the confused taxonomy of New Guinea death adders, and highlights the need for detailed research on this group. We refer our material to *A. antarcticus* simply because that name is in widespread use for all death adders. Venomous, and should be considered dangerous. It was by far the most abundant snake in the area.

Aspidomorphus muelleri (Schlegel, 1837)

Wet forest; terrestrial; probably a sit and wait predator; presumably small vertebrates; dorsum mostly dark brown to black, often with gold reticulations on the head, venter white to cream; SVL 400 - 600 mm; probably oviparous; clutch size unknown; known from throughout New Guinea and associated islands and the Bismarck Archipelago, also Ceram, at elevations from sea level to 1500 m. Venomous, but generally not considered dangerous.

Micropechis ikaheka (Lesson, 1829)

Mostly wet forest but also in damp disturbed areas; terrestrial diurnal and nocturnal, semi-fossorial; probably a sit and wait predator; invertebrates such as earthworms and small vertebrates; coloration variable, dorsum mostly reddish brown to cream with indistinct brown bands, these darker posteriorly, head dark brown to black, venter cream to dull yellow; SVL 1000-1400 mm; oviparous; clutch size unknown; recorded from throughout New Guinea and associated islands, including the Aru group at elevations from sea level to 1500 m. Venomous, and should be considered dangerous.

Toxicocalamus (near *loriae*)

Mostly wet forest but also in damp disturbed areas; terrestrial, semi-fossorial; probably a sit and wait predator; earthworms and insects; dorsum olive brown to dark brown, venter cream to yellow; SVL 500-600 mm; oviparous; clutch unknown; *Toxicocalamus loriae* is known from scattered localities throughout New Guinea and associated islands, at elevations from sea level to 1500 m. Venomous, but not considered dangerous.

Birds Recorded in the Lakekamu Basin

Birds recorded within the Lakekamu Basin (Beehler *et al.*1995) and on the 1996 RAP at the Ivimka Research Station. The taxonomic sequence and organization of major taxa (families and sub-families) follows Sibley and Monroe (1990). However, generic and species level names used in standard PNG references (Beehler *et al.* 1986 and Coates 1985, 1990) are retained where they differ from Sibley and Monroe (1990) in order to make this list more utilitarian. Species designated with an X in the first column were recorded on the RAP survey; those without an X were recorded elsewhere in the Lakekamu Basin (Beehler *et al.* 1995).

SPECIES	IVIMKA	ABUNDANCE[1]	GUILD[2]	EVIDENCE[3]
CASUARIIDAE				
Casuarius casuarius	X	U	F	T
MEGAPODIIDAE				
Talegalla fuscirostris	X		O	V, T
Megapodius freycinet	X	R	O	S
BUCEROTIDAE				
Rhyticeros plicatus	X	C	F	T
CORACIIDAE				
Eurystomus orientalis			I	
ALCEDINIDAE				
Alcedo azurea	X	C	CI	V
Alcedo pusilla			CI	
Ceyx lepidus	X	A	CI	V
DACELONIDAE				
Dacelo gaudichaud	X	A	CI	T, N
Clytoceyx rex	X	U	CI	S
Halcyon nigrocyanea	X	R	CI	S
Halcyon macleayii			CI	
Halcyon sancta			CI	
Halcyon torotoro	X	A	CI	V, T
Melidora macrorrhina	X	A	CI	N
Tanysiptera galatea	X	C	CI	V
Tanysiptera sylvia			CI	
MEROPIDAE				
Merops ornatus			I	
CUCULIDAE				
Cacomantis variolosus	X	R	I	S
Rhamphomantis megarhynchus			I	

SPECIES	IVIMKA	ABUNDANCE[1]	GUILD[2]	EVIDENCE[3]
Chrysococcyx minutillus			I	
Chrysococcyx meyeri			I	
Caliechthrus leucolophus			FI	
Microdynamis parva	X	R	F	V, T
Eudynamis scolopacea			F	
Scythrops novaehollandiae			F	
CENTROPODIDAE				
Centropus menbeki	X	C	CI	T
PSITTACIDAE				
Chalcopsitta scintillata	X	A	FN	S
Pseudeos fuscata	X	C	FIN	S
Trichoglossus haematodus	X	A	FIN	T
Trichoglossus goldiei	X	C	FN	T
Lorius lory	X	C	FIN	S
Charmosyna wilhelminae	X	R	FN	S
Charmosyna placentis	X	U	FN	T
Charmosyna pulchella	X	U	FN	S
Probosiger aterrimus	X	U	G	S
Cacatua galerita	X	C	FGI	T
Micropsitta pusio	X	A	HI	V
Cyclopsitta gulielmiterti	X	C	FG	S
Psittaculirostris desmarestii	X	U	FG	T
Geoffroyus geoffroyi	X	U	FG	S
Eclectus roratus	X	C	FGN	S
Psittrichas fulgidus	X	C	FH	T
APODIDAE				
Collocalia esculenta	X	A	I	S
Collocalia vanikorensis	X	A	I	S
Mearnsia novaeguineae			I	
Hirundapus caudacutus	X	R	I	S
HEMIPROCNIDAE				
Hemiprocne mystacea	X	C	I	S
TYTONIDAE				
Tyto tenebricosa			C	

SPECIES	IVIMKA	ABUNDANCE[1]	GUILD[2]	EVIDENCE[3]
PODARGIDAE				
Podargus papuensis			CI	
Podargus ocellatus	X	U	CI	S
EUROSTOPODIDAE				
Eurostopodus papuensis	X	R	I	S
CAPRIMULGIDAE				
Caprimulgus macrurus			I	
COLUMBIDAE				
Macropygia amboinensis	X	C	FG	S
Macropygia nigrirostris	X	U	FG	S
Reinwardtoena reinwardtii	X	U	FG	T
Chalcophaps stephani	X	C	FG	V
Henicophaps albifrons	X	U	FGI	S
Gallicolumba rufigula	X	U	FG	V
Trugon terrestris	X	U	FG	S
Otidiphaps nobilis	X	R	FGI	S
Ptilinopus magnificus	X	A	F	S
Ptilinopus perlatus	X	C	F	T
Ptilinopus ornatus	X	C	F	S
Ptilinopus superbus	X	C	F	S
Ptilinopus coronulatus	X	R	F	V
Ptilinopus pulchellus	X	A	F	V, T
Ptilinopus rivoli	X	R	F	S
Ptilinopus iozonus			F	
Ptilinopus nanus			F	
Ducula rufigaster	X	C	F	T
Ducula pinon	X	A	F	T
Ducula zoeae	X	U	F	S
Gymnophaps albertisii	X	C	F	S
Goura scheepmakeri	X	C	FGI	S
RALLIDAE				
Rallina tricolor			I	
Rallus philippensis			GI	

SPECIES	IVIMKA	ABUNDANCE[1]	GUILD[2]	EVIDENCE[3]
Eulabeornis plumbeiventris			I	
Amaurornis olivaceus			GHI	
SCOLOPACIDAE				
Tringa hypoleucos			I	
CHARADRIIDAE				
Charadrius dubius			I	
ACCIPITRIDAE				
Pandion haliaetus			C	
Aviceda subcristata	X	R	O	S
Henicopernis longicauda	X	U	CI	S
Macheiramphus alcinus	X	R	C	S
Haliastur indus	X	U	C	S
Accipiter novaehollandiae	X	U	C	S
Accipiter poliocephalus			C	
Accipiter cirrocephalus			C	
Harpyopsis novaeguineae	X	R	C	S
Aquila gurneyi			C	
Hieraaetus morphnoides	X	U	C	S
FALCONIDAE				
Falco peregrinus			C	
ANHINGIDAE				
Anhinga melanogaster			C	
PHALACROCORACIDAE				
Phalacrocorax sulcirostris			C	
Phalacrocorax melanoleucos			C	
ARDEIDAE				
Ardea sumatrana			C	
Egretta alba			C	
Egretta intermedia			C	
Zonerodius heliosylus	X	R	CI	S
Ixobrychus flavicollis			CI	
PITTIDAE				
Pitta erythrogaster	X	C	I	N
Pitta sordida			I	

SPECIES	IVIMKA	ABUNDANCE[1]	GUILD[2]	EVIDENCE[3]
PTILONORHYNCHIDAE				
Ailuroedus buccoides	X	C	O	T, N
MALURIDAE				
Malurus alboscapulatus			I	
Malurus cyanocephalus	X	R	S	
MELIPHAGIDAE				
Myzomela eques	X	R	IN	V
Myzomela nigrita	X	R	IN	N
Myzomela cruentata			IN	
Timeliopsis griseigula	X	C	IN	S
Melilestes megarhynchus	X	C	FIN	V
Glycichaera fallax			IN	
Meliphaga albonotata			FIN	
Meliphaga aruensis	X	A	FIN	V
Meliphaga analoga	X	A	FIN	N
Lichenostomus obscurus	X	A	IN	V, T
Xanthotis flaviventer	X	C	FIN	V, T
Pycnopygius ixioides			IN	
Pycnopygius stictocephalus	X	C	FIN	S
Philemon meyeri	X	A	FIN	V
Philemon buceroides	X	A	FIN	V
PARDOLOTIDAE				
Acanthizinae				
Crateroscelis murina	X	C	I	V, T
Sericornis spilodera	X	R	I	N
Gerygone chloronotus	X	C	I	S
Gerygone palpebrosa	X	C	S	
Gerygone chrysogaster	X	A	I	N
Gerygone magnirostris			I	
EOPSALTRIIDAE				
Monachella muelleriana	X	U	I	S
Microeca flavovirescens	X	A	I	V, T
Poecilodryas hypoleuca	X	C	I	T, N

SPECIES	IVIMKA	ABUNDANCE[1]	GUILD[2]	EVIDENCE[3]
POMATOSTOMIDAE				
Pomatostomus isidorei	X	A	I	V, T
CORVIDAE				
Cinclosomatinae				
Ptilorrhoa caerulescens	X	A	I	V, T
Pachycephalinae				
Pachycephala hyperythra	X	R	I	N
Pachycephala soror			I	
Pachycephala simplex	X	C	I	N
Pachycephala aurea			I	
Colluricincla megarhyncha	X	A	I	V, T
Pitohui kirhocephalus	X	C	FI	T, N
Pitohui ferrugineus	X	A	FI	T, N
Pitohui cristatus			FI	
Corvinae				
Corvus tristis	X	A	O	T
Manucodia atra	X	U	FI	N
Manucodia keraudrenii	X	U	FI	T
Ptiloris magnificus	X	C	FI	V, T
Cicinnurus magnificus	X	R	FI	S
Cicinnurus regius	X	A	FI	T
Seleucidis melanoleuca	X	U	FIN	N
Paradisaea raggiana	X	A	FI	S
Cracticus cassicus	X	A	O	T
Cracticus quoyi	X	C	O	N
Peltops blainvillii	X	C	I	S
Oriolus szalayi	X	A	FI	S
Coracina novaehollandiae			FI	
Coracina caeruleogrisea			FI	
Coracina boyeri	X	U	FI	S
Coracina tenuirostris			FI	
Coracina schisticeps	X	C	FI	S
Coracina melas (=melaena)	X	U	FI	S

SPECIES	IVIMKA	ABUNDANCE[1]	GUILD[2]	EVIDENCE[3]
Campochaera sloetii	X	A	F	S
Lalage leucomela			FI	
Dicrurinae				
Rhipidura leucophrys			I	
Rhipidura rufiventris	X	A	I	V, T
Rhipidura threnothorax	X	I	N	
Rhipidura maculipectus			I	
Rhipidura leucothorax			I	
Rhipidura rufidorsa	X	C	I	V
Chaetorhynchus papuensis	X	U	I	S
Dicrurus hottentottus	X	C	FI	V
Monarcha guttulus	X	A	I	V, T
Monarcha manadensis	X	A	I	N
Monarcha chrysomela	X	U	I	V
Arses telescophthalmus	X	C	I	N
Myiagra alecto	X	U	I	S
Machaerirhynchus flaviventer	X	C	I	V, T
MUSCICAPIDAE				
Zoothera dauma			I	
STURNIDAE				
Aplonis metallica	X	A	FIN	S
Aplonis mystacea	X	R	FIN	S
Mino anais	X	U	F	S
Mino dumontii	X	C	F	S
HIRUNDINIDAE				
Hirundo tahitica			I	
CISTICOLIDAE				
Cisticola exilis			I	
NECTARINIIDAE				
Dicaeum pectorale	X	C	FIN	N
Nectarinia aspasia	X	U	IN	S
Nectarinia jugularis			IN	

SPECIES	IVIMKA	ABUNDANCE[1]	GUILD[2]	EVIDENCE[3]
MELANOCHARITIDAE				
Melanocharis nigra	X	A	FI	V
Toxorhamphus novaeguineae	X	C	IN	V, T
Oedistoma iliolophus	X	U	IN	N
Oedistoma pygmaeum			IN	
PASSERIDAE				
Erythrura trichroa			G	
Erythrura papuana	X	R	G	N
Lonchura caniceps	X	U	G	S

Includes species seen at Ivimka during scouting trips but not during the RAP survey per se. Abbreviations:

[1]ABUNDANCE:
   A - abundant
   C - common
   U - uncommon
   R - rare

[2]GUILD: Foraging guild, or trophic specialization, follows Bell (1982) with modifications based on more recent literature and personal observation. Combinations of letters indicate diet is commonly a mix of different categories.
   C - carnivore (vertebrate prey)
   F - frugivore
   G - granivore
   H - herbivore
   I - insectivore (including other invertebrate prey)
   O - omnivore
   N - nectarivore

[3]EVIDENCE:
   T - tape recording (primary cut). Recordings are deposited at the Library of Natural Sounds, Cornell laboratory of Ornithology, Ithaca, New York, USA. LNS catalog numbers 79-800 to 79-812 and 79-873 to 79-918.
   V - voucher specimen(s) collected and deposited at PNG National Museum (Port Moresby) or Academy of Natural Sciences (Philadelphia). Specimens are study skins and tissue samples preserved in buffer. Specimens not cataloged at time of publication.
   N - netted and released.
   S - seen and/or heard only.

Mammals of the Lakekamu Basin

Mammal species that could hypothetically occur within the Lakekamu Basin. Under "Evidence" are listed those species encountered during the RAP survey. Species of conservation concern according to Flannery (1995) are listed under "Status" and species of concern according to the IUCN (1996) are listed under "IUCN." Taxa in bold face were recorded during the RAP survey.

FAMILY	GENUS	SPECIES	EVIDENCE	STATUS[1]	IUCN[2]
Tachyglossidae	*Tachyglossus*	*aculeatus*			
	Zaglossus	*bruijnii*		E	T
Dasyuridae	**Antechinus**	**melanurus**	specimen		
	Dasyurus	*albopunctatus*		V	T
	Murexia	*longicaudata*			
	Myoictis	*melas*			
Peroryctidae	*Echymipera*	*kalubu*	**Echymipera** sp. sighted		
		rufescens			
	Peroryctes	*raffrayana*			
Macropodidae	*Dendrolagus*	*spadix*		U	DD
	Dorcopsis	**luctuosa**	sighted		
	Thylogale	*browni*		V	T
Phalangeridae	*Phalanger*	*gymnotis*			DD
		intercastellanus			
	Spilocuscus	**maculatus**	sighted		
Acrobatidae	*Distoechurus*	*pennatus*			
Petauridae	*Dactylopsila*	*trivirgata*			
	Petaurus	*breviceps*			
Pseudocheiridae	*Pseudochirulus*	*canescens*			DD
Muridae	**Hydromys**	**crysogaster**	specimen		
	Leptomys	*elegans*		U	T
	Anisomys	*imitator*			
	Chiruromys	*forbesi*			
		vates			
	Lorentzimys	**nouhuysi**	specimen		
	Melomys	**leucogaster**	specimen		
		levipes	specimen	U	
		moncktoni			
		platyops	specimen		

FAMILY	GENUS	SPECIES	EVIDENCE	STATUS[1]	IUCN[2]
		rattoides			
		rufescens	specimen		
	Pogonomys	**loriae**	specimen		
		macrourus	specimen		
	Uromys	**caudimaculatus**	specimen		
	Xenuromys	*barbatus*		U	NT
	Rattus	**leucopus**	specimen		
		steini			
	Stenomys	**verecundus**	specimen		
Pteropodidae	*Dobsonia*	*magna*			
		minor			NT
	Macroglossus	*minimus*			
	Nyctimene	**albiventer**	specimen		
		aello		U	NT
		cyclotis		U	NT
	Paranyctimene	**raptor**	specimen		NT
	Pteropus	*conspicillatus*	**Pteropus sp.** sighted		
		macrotis		U	
		neohibernicus			
	Rousettus	*amplexicaudatus*			
	Syconycteris	**australis**	Captured		
Emballonuridae	*Emballonura*	*beccarii*			
		furax		U	T
		nigrescens			
		raffrayana			CT
Hipposideridae	**Aselliscus**	**tricuspidatus**	specimen		
	Hipposideros	**cervinus**	specimen		
		calcaratus	specimen		
		ater			
		diadema			
		maggietaylorae			
		muscinus	specimen	U	T

FAMILY	GENUS	SPECIES	EVIDENCE	STATUS[1]	IUCN[2]
Rhinolophidae	*Rhinolophus*	*euryotis*	specimen		
		megaphyllus			
Vespertilionidae	*Kerivoula*	*muscina*		U	T
	Miniopterus	*australis*			
		macrocneme			
		magnater		U	
		medius			
		propitristis			
		schreibersii			NT
	Myotis	*adversus*			
	Nyctophilus	*microtis*		U	
	Phoniscus	*papuensis*		V	
	Pipistrellus	*wattsi*	specimen	U	NT
		papuanus			NT
		angulatus			
Molossidae	*Chaerephon*	*jobensis*			
	Mormopterus	*beccarii*		U	
	Otomops	*papuensis*		U	T
		secundus		U	T

[1]STATUS (Flannery 1995):

  E - Endangered

  U - Unknown

  V - Vulnerable

[2]STATUS (IUCN 1996):

  CD - Conservation Dependent

  DD - Data Deficient

  NT - Near Threatened

  T - Threatened

Mammalian Reproductive Activity During the Survey

Documented cases of female reproductive activity; the numbers in parentheses are the number of females displaying the activity and the percentage of all adult females captured that this represents. In all cases lactating and pregnant females represent different individuals (if a female was both pregnant and lactating she appears as pregnant only in the table).

TAXA	REPRODUCTIVE ACTIVITY	DATES (1996)
Lorentzimys nouhuysi	pregnant (2, 100%, 2 embryos each with one in each uterine horn)	23 Oct, 29 Oct
Melomys leucogaster	lactating (1, 100%)	26 Nov
Melomys levipes	lactating (2, 33%); pregnant (2, 33%, one with 2 embryos with one in each uterine horn, one with 1 or 2 embryos in one horn)	20 Oct (lact), 26 Oct (lact), 24 Nov (preg)
Melomys platyops	lactating (2, 100%)	11 Nov, 25 Nov
Melomys rufescens	lactating (1, 25%); pregnant (1, 25%, 2 embryos with one in each uterine horn)	25 Oct (preg), 11 Nov (lact)
Pogonomys macrourus	lactating (1, 100%)	25 Nov
Rattus leucopus	lactating (4, 21%, one with 5 and one with 4 hairless young in trap with her, one with one fully-haired eyes-opened young in trap with her)	29 Oct and 22 Nov (w/ hairless young), 13 Nov (with haired young), 25 Nov
Uromys caudimaculatus	pregnant (1, 50%, 2 embryos with one in each uterine horn)	30 Oct
Syconycteris australis	pregnant (3, 100%, one embryo each)	28 Oct, 5 Nov, 16 Nov
Rhinolophus euryotis	lactating (1, 33%); pregnant (1, 33%, one embryo)	5 Nov (lact), 5 Nov (preg)
Aselliscus tricuspidatus	lactating (1, 20%)	29 Oct
Hipposideros diadema	lactating (1, 100%)	28 Oct

Local Names for Landowners in the Lakekamu Basin

GROUP	ALTERNATE NAME	SOURCE
Biaru	Gorua	in Morobe Province
Biaru	Talep	by the Kurija
Biaru	Namusa	by the Kamea
Biaru	Apunge	by the Kovio
Kamea	Kukukuku	by Colonial administrators
Kamea	Watut	in Morobe Province
Kamea	Kobun	by the Biaru
Kamea	Anut	by the Kurija
Kamea	Anikaugi	by the Kovio
Kurija	Kunimaipa	belonging to the Kunimaipa people
Kurija	Goilala	from Goilala district
Kurija	Wanai	by the Biaru
Kurija	Kauiauwe	by the Kamea
Kurija	Apeape	by the Kamea
Kurija	Bobori	by the Kovio
Kovio	Uvan	by the Biaru
Kovio	Koragol	by the Kurija

Population Figures for the Lakekamu Basin

Local population figures (citizens only) in the Lakekamu Basin. Data are from the PNG National Statistical Office 1993.*

GROUP	LOCATION	# HOUSEHOLDS	TOTAL POPULATION	# MALES	# FEMALES
Biaru	Kakoro	27	121	76	45
Kurija	Mirimas	19	99	48	51
Kurija	Totai	14	59	36	23
Kovio	Okavai	26	175	95	80
Kovio	Urulau	28	162	79	83
Kamea	Tekadu	29	157	85	72

* Note: it is difficult to connect the census unit used by the government to geographical names currently in use. Figures were not available for Iruki, Nukeva and Ungima.

Major Biaru clans in the Lakekamu Basin

CLAN	LAND CLAIMS	HEADMAN
Elka	north of Kakoro and west to the Avi Avi River	Kuskom
Kinggari	south of the mountains, east of the Biaru River	Yinip Assi
Kurumu	south along Biaru River	Namun
Kuinauk	along the Si and Nagore Rivers	Yinip Assi

Kurija Lineages in the Lakekamu Basin

LINEAGE NAME	HEADMAN	# ADULT MEN	RESIDENCE
Kurija-Lairao	Joe Dumoi	9	Mirimas
Kurija-Elop	Laim	2	Mirimas
Kurija-Kabeja	Jarau	3	Totai
Kurija-Remai	Jeven	4	Totai
Kurija-Jagwi	Koiem	4	Totai/Mirimas
Kurija-Kepara	Heremai	3	Mirimas
Kurija		5	Mirimas and Port Moresby
Kurija-Ponpon		defunct	
Kurija Enari		defunct	

Recent Marriages - Mirimas Village

GENDER	SPOUSE'S ORIGIN	CURRENT RESIDENCE	BRIDEWEALTH
female	Goilala	Central Province	K800
female	Chimbu	Chimbu Province	K700
female	Sepik	Lae	K5,000* unpaid
male	Mirimas	Mirimas	none/unknown
male	Gereina, Morobe Prov.	Mirimas	none/unknown
male	Mirimas	Mirimas	K600
male	Mirimas	Mirimas	K1,500 unpaid
male	Mirimas	Mirimas	K1,200 unpaid
male	Mirimas	Mirimas	K1,000 unpaid
male	Mirimas	Mirimas	K600

* Or less if they return to the area

Main Kamea lineages in Iruki Village

LINEAGE NAME	POPULATION SIZE
Nautia	15
Apea	1
Kapete	10
Titama	2
Iuta	6
Iwea	2
Yausea	4
Misea	1
Tomte	3
Amdia	1

Recent Marriages - People Now Living in Iruki Village

GENDER	PLACE OF ORIGIN	SPOUSE'S ORIGIN	BRIDEWEALTH
female	Iruki	Aseki	K700 unpaid
male	Ieva	Kamena	none/unknown
male	Kamena	Kamena	K500
female	Iruki	Kuidinga, Gulf Prov.	K1700
female	Tekadu	Kenabia	none/unknown
male	Kamena	Kamena	K220
male	Kamena	Tekadu	none/unknown
male	Finchafen, Madang Prov.	Biaru	none/unknown

Main Kovio lineages of Okavai village

LINEAGE	HEADMAN
Unga	Kama (lineage of Paul Apio, former Premier of Gulf Province)
Upua	Biwi (lives in Kerema)
Ungima	Agavai (works at aidpost in Okavai)
Kongopu	Joseph Mangabe (lawyer in Kerema)
Birabira	Miva (resides in Okavai)

NOTES

NOTES

NOTES